Designing Environments

Series Editors

Michael U. Hensel, Head of Department of Digital Architecture and Planning
Vienna University of Technology
Vienna, Germany

Claudia R. Binder, Dean of EPFL ENAC
École polytechnique fédérale de Lausanne
Lausanne, Switzerland

Defne Sunguroğlu Hensel, Associate Professor, School of Architecture
Southeast University
Nanjing, Jiangsu, China

This series seeks to address the unfolding climate, environmental and ecological crisis from a broad interdisciplinary perspective and in relation to the impact of human transformations of the environment. The aim is to shift away from segregated modifications of the environment divided into domains and scales, systems and objects, with at best minimum negative impact for the environment, towards an integrative interdisciplinary approach that understands, models and modifies the environment in comprehensive and integrative manner and with net positive impact on the environment. This endeavour involves earth, environmental and life sciences, environmental informatics, computer science, and the disciplines that centrally concern the transformation of the terrestrial environments, such as architecture, landscape architecture and urban design, as well as agriculture and food production. From a methodological perspective, computer and data science play a role in facilitating multi-domain and multi-scale models of environments with the purpose of both analysis and design. At the same time, the series will place the discussion in a necessary cultural context and also discuss the need for ethics in which an alternative approach to the transformation of the environment needs to be based on.

The Series Editors welcome book proposals on the following topics: paradigms, theory and methods for integrative inter- and transdisciplinary approaches to understanding, modelling and modifying environments; relevant historical and contemporary case studies; relevant current research projects; related data science and computer science approaches.

Alessandra Battisti • Michael U. Hensel
Defne Sunguroğlu Hensel
Editors

Urban-Rural Dialogue in Green City Design

Advancing Urban Social-ecological Systems

Springer

Editors
Alessandra Battisti
Department of Planning, Design and Technology of Architecture (PDTA), Faculty of Architecture
Sapienza University of Rome
Rome, Italy

Michael U. Hensel
Department for Digital Architecture and Planning, Faculty for Architecture and Planning
Vienna University of Technology
Vienna, Austria

Defne Sunguroğlu Hensel
Architecture Internationalization Demonstration School
Southeast University
Nanjing, Jiangsu, China

ISSN 2730-6526 ISSN 2730-6534 (electronic)
Designing Environments
ISBN 978-3-031-88317-0 ISBN 978-3-031-88318-7 (eBook)
https://doi.org/10.1007/978-3-031-88318-7

This Springer imprint is published by the registered company Springer Nature Switzerland AG
The registered company address is: Gewerbestrasse 11, 6330 Cham, Switzerland

To our children.

Preface

This forth volume in the *Designing Environments* book series, evolved from the 6th International Architecture and Environment Symposium with the title "Cities, Green Systems, Human Health and Well-being," which took place online in 2021 during the peak of the COVID pandemic. The symposium was hosted by Sapienza University of Rome, Italy, and co-organized by TU Wien. One of the key topics that emerged from the presentations and discussions was rural-urban dialogue, and the way cities are linked with their natural and agricultural surroundings. This volume presents part of this fast-evolving discussion, bringing together thoughts on and frameworks for planning and design, sustainability and resilience, human health and well-being, nature-based solutions, anthropic dynamics and pressures, ecology and biodiversity, urban footprints and metabolism, urban forests and gardens, agri-food landscapes, food cycles, flows, sheds, etc. Despite the range of topics presented in this volume the topic of rural-urban dialogue requires broader, deeper, and more inclusive debates and research, and shows to be an important aspect en route to sustainable development of urban environments and green cities.

Rome, Italy — Alessandra Battisti
Vienna, Austria — Michael U. Hensel
Hamburg, Germany — Defne Sunguroğlu Hensel

Acknowledgments

We extend our sincere gratitude to all the authors for their insightful contributions and to the team at Springer Nature for facilitating an impeccable book production process.

Contents

Editors and Contributors

About the Editors

Alessandra Battisti is an architect and Full Professor of Environmental Design and Technology of Architecture at the Faculty of Architecture, Sapienza University of Rome. She is Director of the II-level Master Program in "Valorization and Enhancement of Small Historical Centres" and serves as faculty board member and teaching staff of the PhD in Environmental Design and Planning. She is an expert consultant for the European Commission, the Italian Ministry of Universities and Research, and the Italian Ministry of Culture.

Michael U. Hensel is an architect and partner of the architectural practice OCEAN Architecture | Environment. He is University Professor at TU Wien where he leads the Department for Digital Architecture and Planning. He has taught for over three decades in renowned schools in The Americas, Asia, Australia, and Europe. He initiated and is Scientific Director of the International Architecture and Environment Symposia and series editor of the *Designing Environments* book series of Springer Nature.

Defne Sunguroğlu Hensel is an architect and partner of the architectural practice OCEAN Architecture | Environment. She is an Associate Professor for Landscape Architecture and Urban Ecology at Southeast University in Nanjing, China, and scientific board member of the Special Interest Group "Knowledge Discovery in Architecture" at the Centre for Artificial Intelligence and Machine Learning at TU Wien. She is a series editor of the *Designing Environments* book series of Springer Nature.

Contributors

Ahmed Abouaiana Department of Architecture, Faculty of Engineering, Sinai University, Ismailia, Egypt

Alessandra Battisti Department of Planning, Design and Technology of Architecture (PDTA), Faculty of Architecture, Sapienza University of Rome, Rome, Italy

Livia Calcagni Department of Planning, Design and Technology of Architecture (PDTA), Faculty of Architecture, Sapienza University of Rome, Rome, Italy

Department of Digital Architecture and Planning, Faculty of Architecture and Planning, Vienna University of Technology, Vienna, Austria

Alberto Calenzo Department of Planning, Design and Technology of Architecture (PDTA), Faculty of Architecture, Sapienza University of Rome, Rome, Italy

Silvia Cioli Zappata Romana, Rome, Italy

Luca D'Eusebio Zappata Romana, Rome, Italy

Marco Giampaoletti Department of Planning, Design, Technology of Architecture, Sapienza University of Rome, Rome, Italy

Sara Favargiotti Department of Civil, Environmental and Mechanical Engineering, University of Trento, Trento, Italy

Michael U. Hensel Head of the Department of Digital Architecture and Planning, TU Wien, Vienna, Austria

Carmela Mariano Department of Planning, Design and Technology of Architecture (PDTA), Faculty of Architecture, Sapienza University of Rome, Rome, Italy

Margherita Meta Department of Planning, Design, Technology of Architecture (PDTA), Faculty of Architecture, Sapienza University of Rome, Rome, Italy

Herbert Natta CNR-IMATI – Institute of Applied Mathematics and Information Technology, Pavia, Italy

Katia Perini Department of Architecture and Design (DAD), University of Genoa, Genoa, Italy

Angelica Pianegonda Department of Civil, Environmental and Mechanical Engineering, University of Trento, Trento, Italy

Enrica Roccotiello Department of Earth, Environment and Life Sciences, University of Genoa, Genoa, Italy

Francesca Rossi Department of Planning, Design, Technology of Architecture (PDTA), Faculty of Architecture, Sapienza University of Rome, Rome, Italy

Defne Sunguroğlu Hensel Architecture Internationalization Demonstration School, Southeast University, Nanjing, China

Fabrizio Tucci Department of Planning, Design, Technology of Architecture, Sapienza University of Rome, Rome, Italy

Maria Valese Department of Planning, Design and Technology of Architecture (PDTA), Faculty of Architecture, Sapienza University of Rome, Rome, Italy

Chapter 1
Introduction to Rural Urban Dialogue in Green City Design

Alessandra Battisti, Defne Sunguroğlu Hensel, and Michael U. Hensel

Abstract This chapter introduces the topic of "Rural Urban Dialogue in Green City Design". Furthermore, this chapter provides an overview over the 15 chapters of the book considering the various presented perspectives and projects related to and motivated by engendering specific forms of rural urban dialogues.

Keywords Rural urban dialogue · Green city design · Sustainability · Urban systems

1.1 Introduction

This volume, the fourth in the *Designing Environments* book series, evolved from discussions in the wake of the sixth International Architecture and Environment Symposium with the title "Cities, Green Systems, Human Health and Well-being", which took place in 2021 at Sapienza University of Rome, Italy. In this context, the topic of rural urban dialogue loomed large, and it became clear that a more focused effort to frame related initiatives, research and projects was needed. This volume constitutes a first step into this direction, bringing together a variety of topics related to planning and design strategy, sustainability and resilience, human health and well-being, the critical zone, nature-based solutions, anthropic dynamics and

A. Battisti (✉)
Department of Planning, Design and Technology of Architecture (PDTA), Faculty of Architecture, Sapienza University of Rome, Rome, Italy
e-mail: alessandra.battisti@uniroma1.it

D. Sunguroğlu Hensel
Architecture Internationalization Demonstration School, Southeast University, Nanjing, China

M. U. Hensel
Department for Digital Architecture and Planning, Faculty for Architecture and Planning, TU Wien, Vienna, Austria
e-mail: michael.hensel@tuwien.ac.at

A. Battisti et al. (eds.), *Urban-Rural Dialogue in Green City Design*, Designing Environments, https://doi.org/10.1007/978-3-031-88318-7_1

pressures, ecology and biodiversity, urban footprints and metabolism, urban forests and gardens, agri-food landscapes, food cycles, flows, and sheds, etc.

Identifying the boundary between what is urban and what is rural entails tracing a tenuous line which obliges us to recast ruralisation and urbanisation as two processes that, while distinct from one another, also prove complementary, at times even subsidiary, interweaving and overlapping along a frontier where the concept of rurality has long been relegated to the intersection of modernity and tradition, suspended between standing as an almost nostalgic symbol of national identities and serving as an instrument for the implementation of various plans of development (Antrop 2004). The reasons behind so much attention are related to a number of considerations, including the fact that the bulk of the world population, from 2008 onward, lives in cities, while the number of inhabitants concentrated in the primary urban areas is projected to reach a total of 5 billion people by 2030 (UN 2024), with this centralising force constituting the factor that has revolutionised models of consumption in terms of both quality and quantity (Perfecto and Vandermeer 2010), raising questions as to the future of food and food supplies (Beddington 2010), in addition to issues of land consumption, pollution and, in absolute terms, climate change.

Considering its astounding growth, the urban system, though it already consists of well-established sites of community life, including some of the highest expressions of the cognitive and planning capabilities of human society, in addition to being endowed with a wealth of knowledge, creativity and capacity for innovation, still has need of innovative approaches to activate resources, awareness and an equitable balance (Vandecasteele et al. 2019). The available natural capital could be optimised for the entire world population, a series of geophysical conditions and parameters, including the average temperatures of the earth's surface, the temperatures of ocean bodies, models of precipitation, the oxygen content and acidity of sea water, all of which are changing so rapidly that nature is unable to adjust, leading to threshold levels that are no longer sustainable for many of the earth's ecosystems.

In 2015, the United Nations' global agenda for sustainable development set objectives (Sustainable Development Goals) that regard, land, and soil, among other factors, to be made a part of short- and medium-term national programs slated for fulfilment by 2030:

- guaranteeing that land consumption does not exceed demographic growth.
- guaranteeing universal access to public green spaces which are safe, inclusive, and open to everyone.
- ensuring that land and soil deterioration, is neutral, a cornerstone objective for maintaining ecosystem functions and services.

Rather than providing forecasts of how these changes will take place, or what will be the underlying causes, the chapters presented in this book set out to grasp how, and in which ways, changes are occurring in the social outlooks, values and technologies employed in dealing with the climatic and economic transformations underway, and more specifically how the urban-rural dialogue is modifying our cities. Each chapter identifies specific factors of change which will have significant influence on urban living, such as the growth of the world population; exposure to risks arising from climate change; cyclical economic crises; persistent levels of poverty; waves of immigration; the exponential rise use of digital technologies; the urgent demand on the part of the earth's inhabitants for more equitable, healthier living spaces.

These chapters address relations between the urban and rural sectors, framing the concept of rurality as part of a balanced dialogue with urban areas, delving into how these complex systems and their models of governance operate. The book examines the extent to which recent forms of rural living are replacing urban customs, while also analysing how the urban and rural sectors have jointly begun to form new urban systems: full-fledged ecosystem models whose innovative approach to decision-making has made the lives of their inhabitants more inclusive, based on logics of collective intelligence that produce shared value (Ernstson 2013). A further focus of the book is the role that technology plays within these ecosystems in the decades to come, in order to bring them in line with the policies of the European Union, including objectives of transition in the fields of energy, ecology and digital technology, through the implementation of specific measures under which the habitat, energy production and the supply of food can be integrated with one other, at the same time as systems for the recovery of water and materials provide incentives for sustainable development and practices of circular economics.

The trend of agriculture entering urban spaces comes with numerous cultural ramifications and practical effects, leaving open questions such as how incisive a role urban vegetation/agriculture can play, as well as the potential risks tied to partial or self-referential outlooks. Of note, in perusing the book's chapters, is their analysis of interrelations between projects and contexts, as well as between timeframes and the main topics. Around these ever-changing and noteworthy buffer zones, a varied set of underpinnings have taken shape: laws, political aims, modes of social behaviour, dreams for the future, economic models, investment strategies, moral values, exemplary episodes, market dealings: all elements that, in combination with the production, circulation and exchange of both tangible and intangible goods, transform the areas in question into processes which, in certain instances, separate themselves from the outcomes for which they were put in place, giving rise to a form of productivity that, for established communities, can serve as the driving force behind projects of urban and peri-urban forestation/agriculture. In some cases, this productivity arises even in the absence of a project, frequently generating not mere wealth, but a shared wellbeing available to everyone. These zones of rural dialogue are often found on the margins, in the cracks of forgotten portions of cities which may also host attempts at initiatives of "neighbourhood government", to restore deteriorated public spaces. Efforts to create a methodology for identifying and classifying the potential output of rural transformations of marginal urban areas, taking into consideration the presence, state and possible development of population centres, as well as the type of natural/rural context involved, plus interactions with the local community and the nature of the government administration, become of fundamental importance in formulating models and plans for occupying open spaces subject to deterioration or abandonment while investigating new forms of social vitality, as well as more ethical, ecological approaches to taking advantage of natural resources. The book's analysis of rural paradigms is also closely tied to assessing the countryside not only as an important natural and cultural resource that must be safeguarded, but also a driving force towards the coherent development of marginal urban areas.

Digital technology can provide any number of tools and methodologies for dealing with the complexity of marginal areas, though steps must first be taken to establish a widely supported framework and system of classification capable of clearly defining the working process.

Some of the chapters propose and encourage a comparative, intelligent approach to integrating digital technology with the natural and urban environment. Representations of any type of human activity must be analysed apart from considerations of their complexity, with a clear-sighted perspective or point of view being introduced to obtain a model as close to reality as possible. While the "modelling" of a phenomenon results in an abstraction, a representation of reality, it also serves as an opportunity to depict what is taking place, at the same time as all the attendant relations, connections, and other elements, both visible and invisible, are analysed.

The authors address topics of efficiency involving ecology, energy, and climate, together with, in certain cases, the question of renewing the methodological framework and the operating instruments for reducing climate-altering emissions while optimising the capacity to adapt to climate change, to promote implementation of a cyclical process that fosters resilience. In the 15 chapters, the book analyses both the traditional and the innovative frontiers of the dialogue between what is rural and what is urban, ranging from urban agriculture to gardening, including aquaponic and aeroponic cultivation, with the methods employed being differentiated by the types of devices utilised, the labour required, and the capital invested. These methods cover a wide array of objectives, such as food production, augmentation of biodiversity, promotion of health and nutritional education, plus social inclusion and the creation of business incubators and planning procedures. The goal is to establish an underlying foundation for understanding and encouraging an ever-greater integration of rural realities within the context of cities, identifying shortcomings, and suggesting new planning combinations able to accelerate the enactment of a different approach to the concept of urban renewal.

As such, the debates portrayed in this book can be summarised by the following topics:

- the need for a critical review of relations between sites of production and consumption, between the city and the country, between urban and peri-urban areas, an assessment that is treated in the book not only as a key consideration in policies having to do with food, but as a crucial rediscovery of the importance of the physical and organisational proximity of numerous social, cultural, and economic relations.
- the urgency of arriving at a widespread use of green technologies and approaches within established cities, and especially European methods that, in combination with institutional and political decisions tied to processes of circular economics,

can promote, and bring to maturation, an ecological transition towards climate neutrality.
- the importance of the open question of how to manage climate emergencies, which brings into play the study of new methodologies of research governed by data-driven systems that will make it possible, in the decades to come, to prevent and contrast emergencies tied to global warming, as well as to rising sea levels and air pollution, thanks in part to the formulation of urban-rural scenarios capable of generating innovative typologies of public spaces in cities.
- the onset of a full-fledged culture of urban self-sufficiency, based on community and practices of social empowerment capable of dealing with the complexity of the changes and transformations underway, all while expanding the capacity of shared, collective decision- making to cover more extensive geographic sectors, so that it can potentially serve as an interface for reformulating ties and relations between urban and rural areas.
- the birth of initiatives that serve to promote social inclusion, including therapeutic and/or educational efforts drawing on inclusive practices aimed at the more fragile sectors of society, and which often take a social-welfare approach, so as to reduce disparities in health and wellbeing, in accordance with the individual's right to the concepts of good health and equity, reinforcing awareness of the role of joint action on the part of the public, the third sector, consumers and businesses.

In sum, the book outlines some possible scenarios for the future, each of which necessarily implies any number of institutional and political visions, together with a series of values tied to the very purpose and concept of introducing aspects of rural life into day-to-day urban existence within contexts that, although they can occasionally appear quite distant from one another in terms of considerations of economics, climate and social concerns, can also give rise to a coherent overview, which is required when dealing with a topic that is so complex that resists facile categorisation within any pre- established framework.

1.2 Chapter Overview

This book contains 11 chapters that focus on questions of rural-urban dialogue for green city design from different perspectives.

Authored by the editors Alessandra Battisti, Defne Sunguroğlu Hensel, and Michael Hensel, Chap. 1 introduces the topic and provides an overview over the 16 chapters of the book.

Authored by Michael Hensel, Chap. 2 is entitled "Rural Urban Dialogue as Experience: Urban Hiking and Human Nature Interaction", examines the typology of urban hiking paths as a way of delivering various aspects of the Sustainable Development Goals (SGDs) of the United Nations. Focus is placed on SDG 3 "Good Health", SDG 11 "Sustainable Cities and Communities", and SDG 15 "Life on Land", with respect to urban ecosystem services and human health and well-being in urban environments. The topic of urban hiking as a way of providing human-nature interaction is discussed as an opportunity for experimental designs leading to urban prototypes comprising architecture and landscape architecture hybrid typologies.

Authored by Alessandra Battisti, Maria Valese, and Herbert Natta, Chap. 3 is entitled "Methodology, Strategies, and Actions for the Sustainable Development of Rural Inner Areas" introduces a systemic, multi-scalar, data-driven approach to strategies and actions for the development of rural inner areas. This leads to a methodological framework for managing the complexity of marginal territories, to enhance processes of regeneration. In this context, spatial data supports multi-scalar and multidisciplinary analysis, integrating information on the value components of the natural, tangible, and intangible cultural heritage, along with the wealth of human and social capital available. This leads to an operative, holistic vision for engaging local stakeholders in a participatory co-design process. The chapter shows how co-creation with local stakeholders, and activation of physical and narrative interconnections between selected points of interests (POIs) can enhance accessibility and attractiveness. This approach transcends the perceived urban/natural dichotomy and provides new classification models and initiatives to foster sustainable growth of rural areas by reconnecting them to the social, cultural, and economic networks of their surrounding territories.

Authored by Fabrizio Tucci and Marco Giampaoletti, Chap. 4 is entitled "The Decarbonization of Urban Districts in the Mediterranean Area by 2050: Green and Grey Solutions", addresses climate changes and emissions of greenhouse gases. This chapter outlines strategies, methodologies, and actions to contribute to the decarbonization of urban districts in the Mediterranean area. Focus is placed on: (1) "green" strategies and actions, through the identification and selection of tree species present in the Mediterranean area with high carbon absorption capacities, and (2) "grey" strategies, actions, and solutions, through the adoption, at various scales, of available technologies and processes for carbon storage (in building envelopes, open space pavements, and underground). The aim is to define a new methodological-applicative approach to support design and meta-design phases in the regeneration of urban districts with focus on climate change adaptation and mitigation. The research targets designers, architects, and planners and aims to establish a knowledge base and decision-making support through a framework of best practices and actionable measures in the field of construction.

Authored by Carmela Mariano, Chap. 5 is entitled "Ecological and Landscape Networks as Strategic and Structural Components of Inter-municipal Planning", discusses research on models of government of metropolitan territories and the relations between urban planning and emerging forms of supra-municipal aggregation, with reference to metropolitan areas in which bottom-up initiatives of inter-municipal associationism and supra-municipal government bodies coexist. The chapter investigates the role of connective networks of green and natural and anthropic components within inter-municipal planning processes aimed at a strategy of integrated urban regeneration. European case studies are discussed with the aim to identify stepping stones for an innovative planning system that adopts an inclusive conception of the territory for unitary knowledge and design of the different components. Within this framework, the chapter focuses on Territorial Coherence (SCOT) of the Grenoble Urban Region, in which the system of networks of natural, semi-natural and anthropic components supports the implementation of urban regeneration strategies and metropolitan territorial rebalancing and the development and reconfiguration of the physical and socio-economic assets of the territory.

Authored by Francesca Rossi and Margherita Meta, Chap. 6 is entitled "Cultivating Public Spaces. Methods and Procedures for Inclusive Urban Re-Appropriation Practices". This chapter examines the transformative potential of cultivating public green spaces within economically distressed and marginalised public settlements, which provide opportunities to promote social cohesion, support the circular economy and encourage solidarity-based citizenship practices. The chapter shows that informal cultivation initiatives typically take place in green spaces of various sizes, often municipally owned but managed by private citizens, which organise themselves in specific associations which employ physical revitalisation strategies to reclaim neglected public green spaces within their settlements. Beyond mere economic support, these informal cultivation areas act as a crucial social glue, compensating for the absence of gathering spaces and promoting a conscious use of resources. They nurture a sense of community, encourage responsible use of public spaces, facilitate active citizenship, and foster peaceful interactions among residents. They also implement adaptive solutions, circular economy practices and innovative cultivation techniques, supporting a healthier and resilient lifestyle, and the uncontrolled urbanisation of green areas.

Authored by Katia Perini and Enrica Roccotiello, Chap. 7 is entitled "Urban Regeneration with Nature-Based Solutions: Human and Nature Perspectives". The chapter addresses the topics of urban ecosystems, densely populated areas, anthropic pressure, lack of green spaces, habitat fragmentation and local biodiversity loss, and more specifically the heat-island effect, poor air quality, and the deterioration of water resources and ecosystems. Focus is placed on nature-based solutions that can play a significant role in regeneration processes, by providing ecosystem services for the improvement of urban ecological and environmental quality. The chapter discussed the example of the Genoa (Italy) EU H2020 funded UNaLab pilot project and related monitoring activities, implemented in the field of human wellbeing and support of local biodiversity.

Authored by Sara Favargiotti and Angelica Pianegonda, Chap. 8 is entitled "Regenerative Foodscapes as a Holistic and Integrated Approach for Ecological Transition in the Alpine Territories". The chapter introduces the city of Trento's initiative and the EIT Food research projects, which focus on revitalizing abandoned and underutilized open areas, creating productive and recreational landscapes that promote agroecology. The chapter shows how these efforts begin to establish inclusive, sustainable, and circular habitats, mitigating climate change risks through natural assets, and enhancing socio-ecological and biodiversity values in peri-urban areas. The research focuses on the multifunctionality of urban and peri-urban agriculture in Trento, seeking to provide environmental, economic, and social benefits, limiting urban expansion, improving water management, regulating microclimates, diversifying natural environments, promoting local biodiversity and local food production, providing employment opportunities, and supporting ecosystem services.

Authored by Livia Calcagni and Alberto Calenzo, Chap. 9 is entitled "Growing Urban Health—Questioning the Role of Urban Gardening in Distressed Urban Areas". Urban agriculture, especially community gardens, can significantly enhance the socio-environmental condition of vulnerable districts and communities. The chapter examines urban agriculture and its value beyond profitability and food production in terms of social empowerment, health, and well-being. The paper seeks to uncover the reasons behind the failure of a community garden initiative that took place in a distressed urban area of the city of Rome, as well as to clarify the connections between community garden initiatives, socioeconomic context, and urban structure. Case study research served a comparison of best practices. All case studies are linked to bottom-up processes in marginal areas where public space could represent a catalyst that can embrace cultural, social, environmental, and ecosystemic aspects. The study shows that community garden initiatives must include a community's primary needs, institutional and financial support, extensive engagement of inhabitants and stakeholders.

Authored by Silvia Cioli and Luca D'Eusebio, Chap. 10 is entitled "Cultivating Cities: Urban Gardenings Impacts and Challenges". The chapter highlights the importance of urban gardening, which has become a practice that provides for different needs and promotes a sense of community stimulating social cohesion. Moreover, it strengthens urban ecosystems by increasing urban biodiversity and fighting climate change, it provides models of innovative production and management of public spaces, mitigating spatial and environmental injustice. Furthermore, the chapter shows that the related impact positively affects social, economic, environmental, political aspects of urban life, but requires the right balance between spontaneous and institutionalized initiatives.

Authored by Ahmed Abouaiana, Chap. 11 is entitled "Constructed Wetlands as Green Infrastructure to Improve Quality of Life". The chapter commences from the position that Green Infrastructure (GI) is vital in enhancing the built environments' ecosystems. It provides numerous benefits (environmental, social, climate change mitigation and adaptation, and biodiversity); GI has different artificial features, such as building green roofs, eco-bridges, or constructed wetlands (CWs). CWs are low-cost, nature-based solutions to treat wastewater and improve water quality. Beyond the technical solutions, this chapter highlights the role of multifunction CWs in improving ecosystems and quality of life in urban areas. The chapter describes several case studies with different CW characteristics and scales: limited architectural and landscaping practices integrated with spatial planning and city development.

References

Antrop M (2004) Landscape change and the urbanization process in Europe. Landsc Urban Plan 67:1–4. https://doi.org/10.1016/S0169-2046(03)00026-4

Beddington J (2010) Food security: contributions from science to a new and greener revolution. Philos Trans R Soc Lond Ser B Biol Sci 365:61–71. https://doi.org/10.1098/rstb.2009.0201

Ernstson H (2013) The social production of ecosystem services: a framework for studying environmental justice and ecological complexity in urbanized landscapes. Landsc Urban Plan 109(1):7–17. https://doi.org/10.1016/j.landurbplan.2012.10.005

Perfecto I, Vandermeer J (2010) The agroecological matrix as alternative to the land—sparing/agriculture intensification model. Proc Natl Acad Sci USA 107:5786–5791. https://doi.org/10.1073/pnas.0905455107

United Nations (2015) Transforming our world: the 2030 Agenda for Sustainable Development. Resolution adopted by the General Assembly on 25 September 2015, A/RES/70/1. United Nations, New York. Available online: https://sdgs.un.org/2030agenda

United Nations, Department of Economic and Social Affairs, Population Division (2024) World population prospects 2024, data sources. UN DESA/POP/2024/DC/NO. 11

Vandecasteele I, Baranzelli C, Siragusa A, Aurambout JP (2019) The future of cities—opportunities, challenges, and the way forward. Publications Office of the European Union, Luxembourg. Available online: https://ec.europa.eu/jrc/en/publication/eur-scientific-and-technical-research-reports/future-cities. Accessed 22 December 2024

Chapter 2
Rural Urban Dialogue as Experience: Urban Hiking Trails and Architecture and Landscape Architecture Hybrid Typologies

Michael U. Hensel

Abstract This chapter examines the typology of urban hiking paths as a way of engendering a key experiential aspect of rural-urban dialogue delivering the arguably closely linked Sustainable Development Goals (SGDs) of the United Nations, SDG 11 "Sustainable Cities and Communities", and SDG 3 "Good Health", SDG 13 "Climate Action", and SDG 15 "Life on Land". The research portrayed in this chapter shows that urban hiking paths can engender and be integrated with constructions that fuse architecture and landscape architecture into new green urban prototypes. This is done with respect to human-nature interaction, mitigation of adverse urban climate effects, delivery of urban ecosystem services and promotion of human health and well-being in urban environments.

Keywords Human health and well-being · Human-nature interface · Sustainable development goals · Urban green prototypes · Urban hiking trails

2.1 Introduction

The report "Our Common Future" published in 1987 by The World Commission on Environment and Development addressed the connection between environment and economic development and outlined the urgency of and requirements for sustainable development. In this context, the report defined the latter as "development that meets the needs of the present without compromising the ability of future generations to meet their own needs." Furthermore, the report emphasized "the vital processes carried out by nature, including stabilization of climate, protection of watersheds and soil, preservation of nurseries and breeding grounds [...] Preserving

M. U. Hensel (✉)
Department for Digital Architecture and Planning, Faculty for Architecture and Planning, TU Wien, Vienna, Austria
e-mail: michael.hensel@tuwien.ac.at

A. Battisti et al. (eds.), *Urban-Rural Dialogue in Green City Design*, Designing Environments, https://doi.org/10.1007/978-3-031-88318-7_2

these processes cannot be divorced from conserving the individual species within natural ecosystems. Managing species and ecosystems together is clearly the most rational way to approach the problem." (The World Commission on Environment and Development 1987) To deliver a platform for sustainable development the United Nations developed 17 Sustainable Development Goals (SDGs) with an extensive set of related sub-goals (United Nations Department of Economic and Social Affairs—Sustainable Development—The 17 Goals 2024). Regarding questions of the environmental impact of urbanization on the one hand and urban environments on the other, SDG 11 on "Sustainable Cities and Communities" is of central importance but can arguably not be tackled in isolation, especially concerning questions of urban ecology and the relation between green cities and urban health and well-being. Other SDGs are inextricably linked with SDG 11, such as SDG 3 on "Good Health and Well-being", SDG 13 on "Climate Action", and SDG 15 on "Life on Land". However, when examining recent progress on these SDGs individually and jointly a discouraging picture emerges. The United Nations 2024 report on the global progress towards the Sustainable Development Goals shows mixed progress on SDG 11 on "Sustainable Cities and Communities", largely marginal progress and regression SDG 3 on "Good Health and Well-being", SDG 13 on "Climate Action", and SDG 15 on "Life on Land" (United Nations—Progress towards the Sustainable Development Goals 2024). The recent report by the European Union shows a similar development (EUROSTAT 2023).

While the shortcoming in addressing the close linkages between various SDGs constitute one problem, another can be found in the current classification of land use systems, which do not provide for hybrid land use, and a limited pallet of typologies for green spaces and hybrid typologies that fuse architecture and landscape architecture.

Recent research has shown that a variety of definitions exist for land use and land cover that are often regionally and locally specific (Nedd et al. 2021). Ritchie and Roser distinguished land use categories for agricultural and non-agricultural land cover, in general divided into agricultural areas (permanent crops, temporary crops, fallow, temporary meadows and pastures, permanent meadows and pastures, inland water bodies, forest (primary forest, other naturally regenerated forest, planted forest), and other land (barren land, urban / build-up land) (Our World in Data 2024). Such definitions and categorization does not include hybrid categories such as urban agriculture, or the possibility that planted forest land cover could be achieved in by very green cities. However, novel efforts in *biophilic design* (Beatly 2011; Söderlund 2019) or *nature integrated design* (Weisser et al. 2022; Hensel and Sunguroğlu Hensel 2023, 2024) and *embedded architectures* (Hensel 2022) could precisely achieve and aim for hybrid land use and land cover arrangements.

Annual surveys on the most livable cities across the planet frequently include the aspect of availability and access to green spaces. These include urban green spaces, but also frequently rural and natural areas in close proximity to cities. What is less frequently considered are linkages between urban and rural green areas that are also green. A key element in this context are urban hiking trails that various cities feature. Examples include, for instance, Vienna in Austria, and Essen in Germany.

Fig. 2.1 Screenshot of the official urban hiking website of the city of Vienna, Austria, 2024

Municipal websites of these cities feature descriptions of urban hiking trails (Figs. 2.1 and 2.2).

Two different types of trails can be distinguished, those that link up urban green spaces and areas within cities, and those that link urban, suburban, and peri-urban rural and natural areas. It is the second type that is of particular interest since the character of linked green spaces changes entirely. It is this type of trail that can provide for human experience of rural-urban connection and dialogue and that is the focal type of this chapter, engendering novel thinking about linked green constructions in the form of architecture and landscape architecture hybrids (Hensel 2021, 2022; Hensel and Sunguroğlu Hensel 2023, 2024). The following aspects are key to this discussion: (1) the rights to ground from a human and non-human perspective (Hensel 2019), (2) ecological connectivity and ecosystem services, (3) human experiences (Hensel and Sunguroğlu Hensel 2025), and (4) human health and well-being. These aspects are discussed in this chapter, followed by two experimental projects that aimed at addressing these aspects.

2.2 Human Experiences

The first aspect that hiking relates to is human experience of environments. In this context, the anthropologist Tim Ingold proposed that there exists a relation between becoming knowledgeable, walking, and the experience of ground and weather: "knowledge is grown along a myriad of paths we take as we make our ways through the world … as [we] negotiate a way through a zone of admixture and interchange between the more or less solid substances of the earth and the volatile medium of air. It is in this zone that all terrestrial life is lived" (Ingold 2010). In this context,

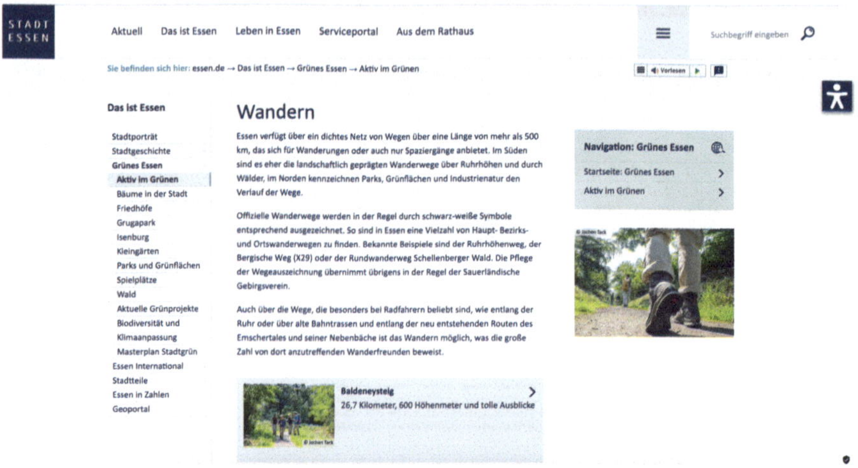

Fig. 2.2 Screenshot of the official urban hiking website of the city of Essen, Germany, 2024

Ingold contrasted two modalities of movement, namely transport and wayfaring: "Transport … carries the passenger *across* a preprepared, planar surface", while the wayfarer "negotiates and improvises a passage as he goes *along* … Along the way, events take place, observations are made, and life unfolds" (Ingold 2010, s126). Ingold associated transport with hard surfaces that seal the ground to enable movement across it, and wayfaring to unsealed ground. For Ingold ground is characterised by four main traits: (1) ground is perceived kinaesthetically in movement; (2) ground is not featureless and level, but instead infinitely variegated; (3) ground is composite in its materiality ("matter from the interweaving of a miscellany of different materials, each with its own peculiar properties" (Ingold 2010, s125)); and (4) ground undergoes continuous transformation (Ingold 2010, s125). It is these four traits that characterise the ground, as well as the emergent path, on which the wayfarer moves. Weather interacts with and shapes the ground, with the wayfarer experiencing both the dynamics of weather and the dynamics of the interaction between weather and ground. In Ingold's view "movement *is* knowing" and enables "a greater sensitivity to cues in the environment and a greater capacity to respond to these cues with judgement and precision" (Ingold 2010, s134) that "increases in concentration and intensity with the fluency of action" (Ingold 2010, s136). And finally, "the ground comprises a domain in which the lives and minds of its human and non-human inhabitants are comprehensively bound up with one another" (Ingold 2010, s135).

The experience of close contact with ground and weather in the "wayfarer" mode of movement therefore constitutes a key experience for humans to interact with and thereby become knowledgeable about locally specific environments.

2.3 Rights to Ground

In order to enable the experience of environment in movement, and by extension dynamic human-nature interactions, the question of rights to access and use of ground necessarily needs to be addressed for both human and non-human actors (Hensel 2019). The topic of hiking trails entails questions of rights to access. Throughout history various provisions that govern the use and temporal appropriation of land owned by others have evolved, for instance the "freedom to roam". There exists evidence of the freedom to roam in many European countries that suggests that such rights might have been common in the past and may have evolved through *practice* over centuries. In Scandinavia such rights are embodies in the "everyman's right". In Norway the 'everyman's right' is an old customary law. Today it is part of the Norwegian Outdoor Recreation Act, which protects the natural basis for outdoor recreation and the public right of access to the countryside. In Sweden the Right of Public Access is part of the Constitution. Understanding the historical and contemporary motivation and customary and common-sense based rights can inform rights to access not only in the countryside, but also in peripheral and urban areas. Regarding urban hiking trails such rights are already implicitly assumed to explicitly defined.

A key question is then how contemporary architectures, settlement pattern and even cities could be designed in accordance with rights to access and use of ground, especially concerning green spaces, whether public or not.

From a non-human perspective this is a more complicated issue. Animals that roam in cities are often viewed as potential hazards. However, this is arguably an outcome for not taking local animal species as important part of local ecosystems into consideration when planning cities and designing neighbourhoods, architectures, and green spaces. Clearly the borders of the territories occupied by animal species do not coincide the plot borders, construction, and infrastructural arteries. This leads to severely constrained spaces from the perspective of animal movement and implies frequently unfortunately also different kinds of accidents that arise from human and animal encounters in spatially constrained situations. However, when approached from the perspective of mitigating adverse climate effects and the relation between ecosystem services and human health and well-being it is useful to make provisions for urban ecosystems.

2.4 Ecosystem Services & Human Health

Gretchen Daily defined ecosystem services as "the conditions and processes through which natural ecosystems, and species that make them up, sustain and fulfil human life [...] Ecosystem services are generated by a complex of natural cycles, driven by solar energy, that constitute the workings of the biosphere—the thin layer near the Earth's surface that contains all life" (Daily 1997, pp. 3–4). According to Daily ecosystem services include "purification of air and water, mitigation of floods and droughts, detoxification and decomposition of wastes, generation and renewal of soil and soil fertility, pollination of crops and natural vegetation, control of the vast majority of potential agricultural pests, dispersal of seeds and translocation of nutrients, maintenance of biodiversity (from which humanity has derived key elements of its agricultural, medicinal, and industrial enterprise), protection for the sun's harmful ultraviolet rays, partial stabilisation of climate, moderation of temperature extremes and the force of winds and waves, support for diverse human cultures, [and providing of aesthetic beauty and intellectual stimulation that lift the human spirit." (Daily 1997, p. 4) A growing amount of research further elaborated the relation between ecosystem services and human health and well-being. Lindgren and Elmquist explained that "research on health effects of ecosystem services have until recently mostly focused on beneficial effects of physical and mental health from spending time in nature or having access to urban green space. However, nearly all of the different ecosystem services mas have impacts of health, either directly or indirectly. Ecosystem services can be divided into provisioning services that provide food and water; regulating services that provide, for example, clean air, moderate extreme events, and regulate the local climate; supporting services that help maintain biodiversity and infectious disease control; and cultural services." (Lindgren and Elmquist 2017) Chiabai et al. pointed out that human "contact with nature and ecosystem services (ES) have been shown to contribute to improved immune system functioning, mood and concentration, while reducing stress and increasing the benefits of physical exercise, with consequent expected reductions in the occurrence of non-communicable diseases (NCDs)." (Chiabai et al. 2018).

However, in urban contexts green spaces are frequently sparsely spread across the city, presenting a clear problem regarding the required connectivity for these green spaces that are often not adequality planned and designed from an ecological perspective and that result in isolated islands that are difficult to reach many species. Ecological connectivity is therefore an important factor for the functioning healthy urban ecosystems. Egerer and Anderson (2020) stated that "urban landscapes, [...] are highly dynamic in land use and often in social composition. Thus, landscape connectivity, especially in cities, must be thought of in a social-ecological framework. This is relevant when considering ecosystem services—the benefits that people derive from ecological processes and properties. As relevant actors move through a connected landscape-scale network, particular services may 'flow' better across space and time." And furthermore, "to integrate the concepts of networks, connectivity, dynamic landscapes, and movement into an ecosystem services framework, we must consider the landscape from a social-ecological perspective because both social and biophysical landscape characteristics influence who experiences which ecosystem services and to what extent through time (Pastur et al. 2016; Bürgi et al. 2015) Additionally, considering the social-ecological connectivity of ecosystem services throughout a network can inform how landscape-scale networks can support human wellbeing and ecological resilience (e.g., resistance to regime shifts)."

In this context, urban hiking trails can combine the provision of human experience, human nature interaction, opportunity for biodiversity and local ecosystem enhancement, provision of ecosystem services and ecological or landscape connectivity by connecting existing and new green spaces along rural-urban gradients. The potential of the diverse yet linked benefits of this type of green space should foster increased interest and research, as well as eventual implementation of urban hiking trails and networks in every city. To transport this vision, it is necessary that urban planners, architects, landscape architects and ecologists co-develop projects that can trigger public interest and demand and municipal governance action.

2.5 Experimental Projects

Two experimental projects by OCEAN net merge architecture and landscape architecture to develop novel green prototypes. In so doing, these projects provide nexuses for hiking trails in which moving through the amalgamation of landscape and construction heightens the experience of movement through environment: the Langley Vale Visitor Hub in Surry, UK, and the Abu Dhabi Flamingo Visitor Centre at the perimeter of the Al Wathba Wetland Reserve (Hensel and Sunguroğlu Hensel 2023, 2024, 2025).

The site of the Langley Vale Visitor Hub is in Surrey, UK. Langley Vale Wood combines ancient woodland, grassland and woodland creation that offer different habitats for local wildlife. The scheme forms an X-shape in plan with the two west-facing volumes rising above ground-level, projecting out to the north and the south respectively. The two east-facing volumes are partially covered by new terrain that slopes down from the roofscape to the east (Fig. 2.3).

Fig. 2.3 Rendered aerial view of the Langley Vale Visitor Centre in Surry, UK. OCEAN net 2016

The west-facing elevation provides a more conventional architectural façade with the massing formed by the two cantilevering volumes that rise from the centre of the hub outwards. Parts of the cantilevering volume feature spaces for various types of birds and bats. The eastern elevation consists of the densely vegetated slope that provides resources for local animal species (Fig. 2.4).

The southern and northern elevations are partly masked by vegetation and reveal only the end faces of the cantilevering volumes (Fig. 2.5).

Arriving through a dense network of hedges and tree stands the visitor approaches the hub from the south-western corner. Entering the structure below the south-western cantilevering volume, the visitor arrives to the central space of the hub that also forms its crossroads from where different paths are available (Fig. 2.6). From here the visitor can choose to ascend the ramped stairway inside the north-western cantilever and to exit from there onto the roof where the hard surface of the interior turns into the soft ground of the roofscape. The green roof features an ecological transition from grassland to forest, with the latter occupying the Eastern slope. From here the visitor can enter back into the building and descend the ramped stairway of the north-eastern volume back into the interior of the hub. The continuous path into, out of, and through the visitor hub is articulated as a variegated landscape experience of changing materiality, enclosure, and exposure to climate and surroundings. The contrasting experiences along this path enhance the creation of greater sensitivity to the specific traits of the local environment and the different species that populate it.

The site of the Abu Dhabi Flamingo Visitor Centre is located at the northern perimeter of the Al Wathba Wetland Reserve (Figs. 2.7, 2.8 and 2.9).

Fig. 2.4 Rendered view of the Langley Vale Visitor Centre in Surry, UK. OCEAN net 2016

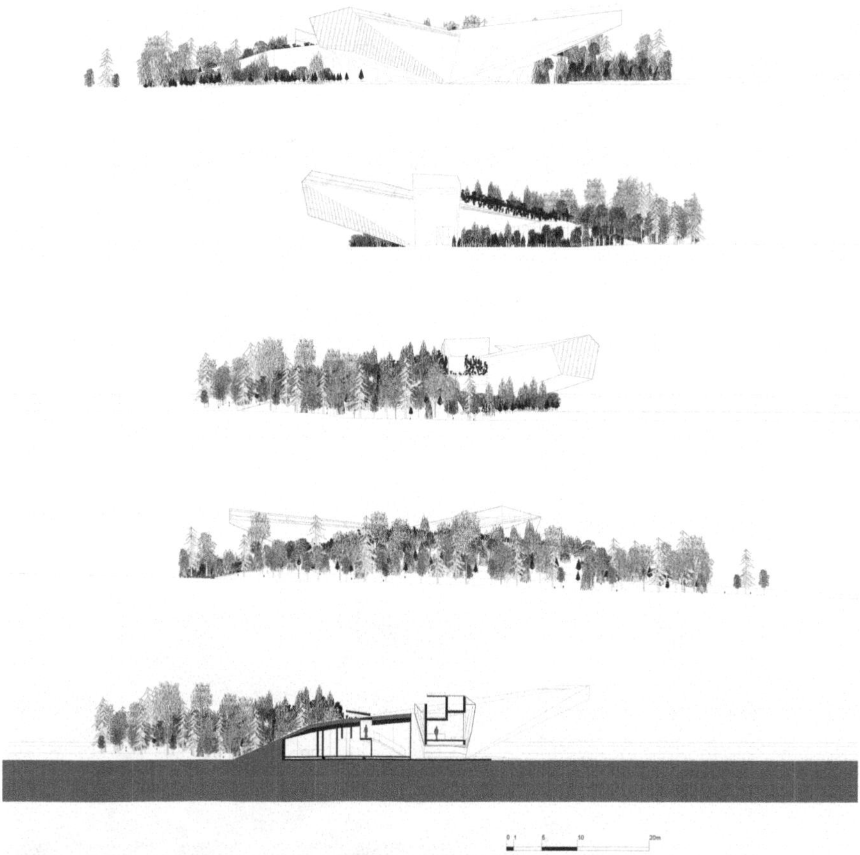

Fig. 2.5 Elevations and cross-section of the Langley Vale Visitor Centre in Surry, UK. OCEAN net 2016

The reserve combines natural, and human made water bodies and is home to Great Flamingos, 260 migratory and non-migratory bird species, and numerous mammal, reptile, insect, and plant species. In the reserve there exist several walking trails for visitors. The scheme comprises of a universally accessible landscape that constitutes a starting point and crossroad for the walking trails. The project articulates an ecological transition zone from the afforested area to the wetland ecosystem (Fig. 2.10).

The scheme consists of a sequence of linked landscape, spatial and microclimatic experiences characterize the scheme: tree groves, the transitional landscape between the afforested areas and the wetlands, the spatial transition between tree groves, sun-sails, and the cavernous interior. Different paths across the site and through the projects are provided and connect to the wetland hiking trails (Fig. 2.11).

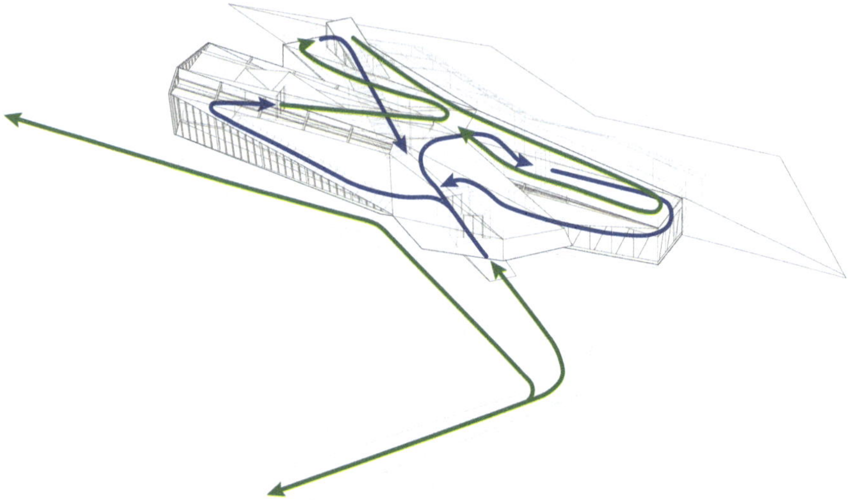

Fig. 2.6 Axonometric view and circulation diagram of the Langley Vale Visitor Centre in Surry, UK. OCEAN net 2016

Fig. 2.7 Rendered aerial view of the Abi Dhabi Flamingo Visitor Centre at the Al Wathba Wetland Reserve. OCEAN net 2020

From the eastern and western directions, the visitor experiences a landscape that gradually transitions from tree groves to the north to the wetlands to the south. The tree groves are composed of local tree species selected for the afforestation of the area. From the northern tree grove, the visitor can enter the interior or ascend to the landscaped roof and from there descend again towards the wetlands. Along this

Fig. 2.8 Rendered view of the Abi Dhabi Flamingo Visitor Centre at the Al Wathba Wetland Reserve. OCEAN net 2020

route only few architectural elements are visible to the visitor, such as the northern entrance. Alternatively, it is possible to walk around the rising landscape to enter the visitor centre from the west, where the western perimeter presents a clear architectural elevation that is not diffused by tree groves or landscape elements. The eastern perimeter of the project is largely masked by a dense line of trees. On the southern elevation architectural and landscape elements alternate. The massive structure that forms the cavernous interior space bridges across the interior along its north-south axis and supports the soil of varying depth to provide for the plants of the roofscape.

In project like this, the question of ecological strategy is of key importance. While the Al Wathba Wetland may have at least in part emerged from human action, its ecosystem may well be competing with the planned afforestation of the area. This competition may entail that either one of the ecosystems might cease to exist in future. Longer term consequences might be understood through analytical models and simulations and stipulations may arise from this as to whether to support one outcome or another. At the spatial scale of the project a transitional area can be demonstrated, while the temporal scale of change may elude design considerations focused on the object alone. Here the experience of the wayfarer leading to "a greater sensitivity to cues in the environment and a greater capacity to respond to these cues with judgement and precision" may offer an inroad but begs the question as to which experience should be provided and how.

In general, these types of experimental projects can serve as a way of developing novel urban green prototypes that can tie in with existing urban green spaces and linear elements such as urban hiking trails, intensifying experiences for human actors and making vital provisions for urban ecosystems and ecosystem connectivity, and thereby for human health and well-being.

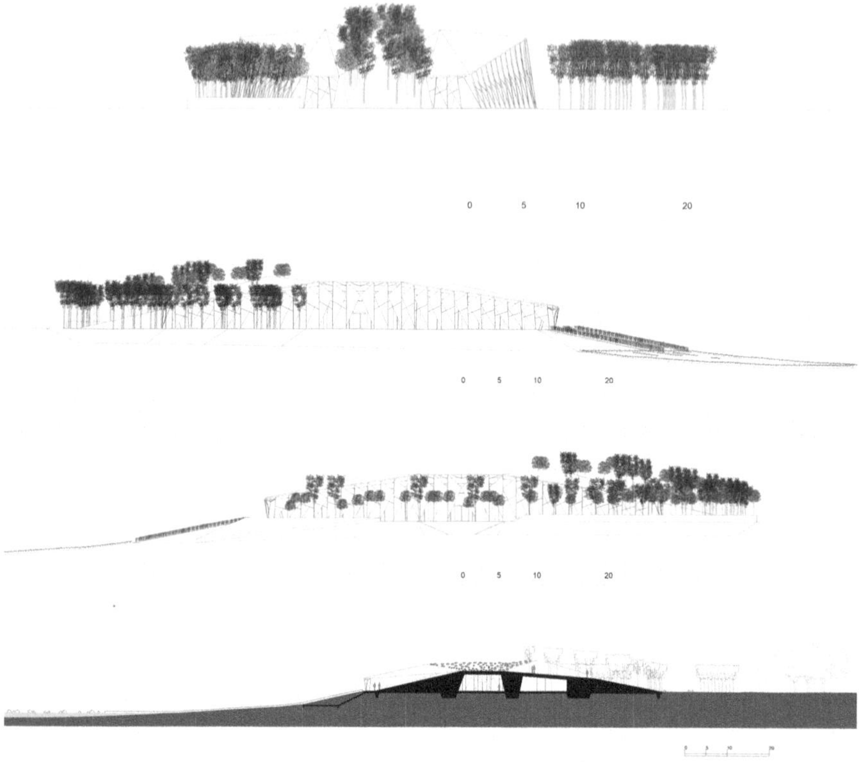

Fig. 2.9 Elevations and longitudinal section of the Abi Dhabi Flamingo Visitor Centre at the Al Wathba Wetland Reserve. OCEAN net 2020

2.6 Conclusion

This chapter set out to show that urban hiking trails can constitute a significant means to promote human-nature interfaces and experiences in the context of urban environments and their surroundings. More specifically this can entail mitigation of adverse urban climate effects, delivery of urban ecosystem services and promotion of human health and well-being in urban environments. This approach is grounded in the realisation that the related UN Sustainable Development Goals are closely linked and need to be considered accordingly. This involves SDG 11 "Sustainable Cities and Communities", SDG 3 "Good Health", SDG 13 "Climate Action", and SDG 15 "Life on Land". Furthermore, this chapter introduced two experimental projects that constitute an intensification and expansion of urban based on types of constructions that fuse architecture and landscape architecture, thereby incorporating a considerable amount of soil and providing territory for plants and animals. In so doing, urban hiking paths in conjunction with these novel types of projects can serve to connect urban and rural green spaces, thereby enabling ecological connectivity in urban environments and extending shared multi-species territories. The

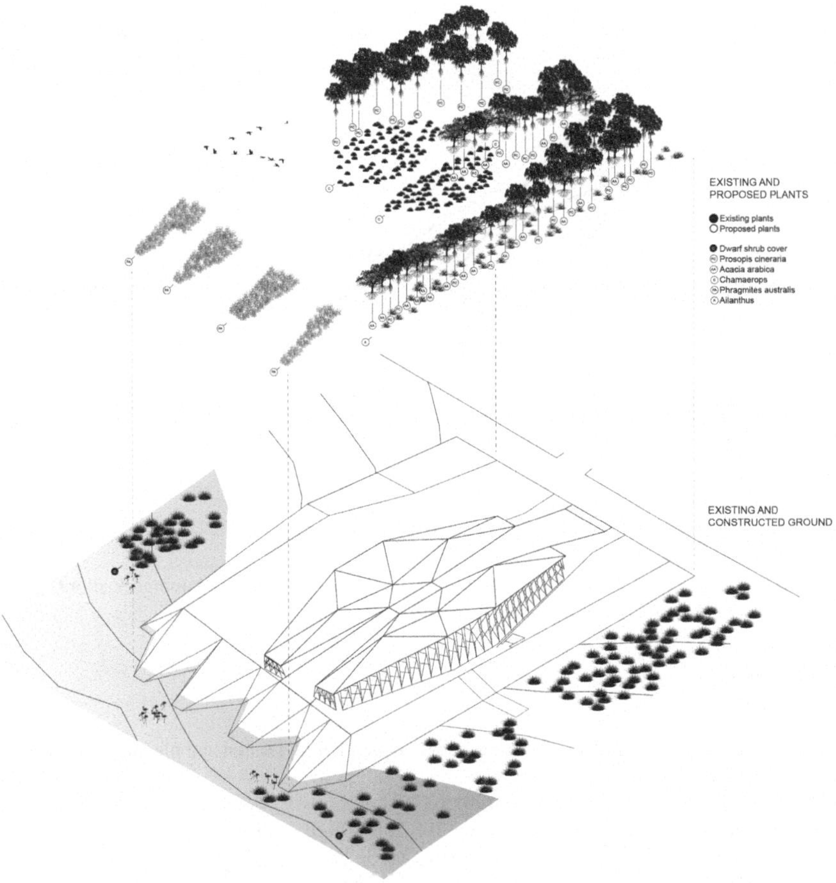

Fig. 2.10 Exploded axonometric view terrain form (bottom) and the plant selection for the Abi Dhabi Flamingo Visitor Centre at the Al Wathba Wetland Reserve. OCEAN net 2020

advantage of this approach is that construction and green spaces are no longer treated as essentially mutually exclusive and that any space in urban environments can be articulated accordingly. This commences with linear connections in the form of hiking trails thereby reducing the required spaces for connections, while utilising the built fabric as the extended ecological and experiential territory. Of central importance is the linking of rural and urban areas that can be effectively enacted in this manner.

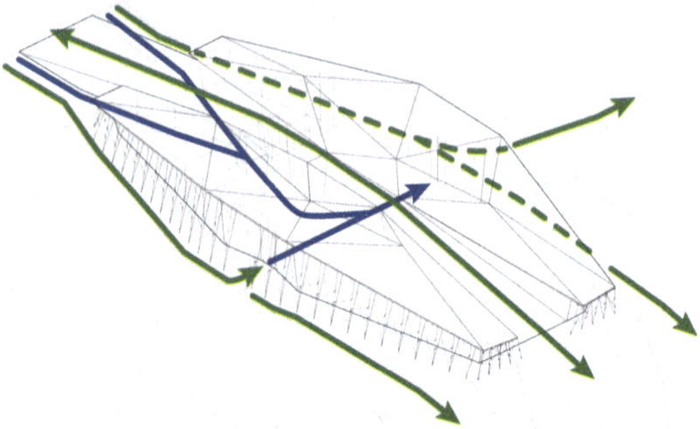

Fig. 2.11 Axonometric view and circulation diagram of the Abi Dhabi Flamingo Visitor Centre at the Al Wathba Wetland Reserve. OCEAN net 2020

References

Beatly T (2011) Biophilic cities—integrating nature and Urban Design and planning. Island Press, Washington

Bürgi M, Silbernagel J, Wu J, Kienast F (2015) Linking ecosystem services with landscape history. Landsc Ecol 30:11–20. https://doi.org/10.1007/s10908-014-0102-3

Chiabai A, Quiroga S, Martinez-Juarez P, Higgins S, Taylor T (2018) The Nexus between climate change, ecosystem services and human health: towards a conceptual framework. Sci Total Environ 635:1191–1204. https://doi.org/10.1016/j.scitotenv.2018.03.323

Daily GC (1997) Introduction: what are ecosystem services. In: Daily GC (ed) Nature's services: societal dependence on natural ecosystems. Island Press, Washington, DC, pp 1–10

Egerer M, Anderson E (2020) Social-ecological connectivity to understand ecosystem service provision across networks in urban landscapes. Land 9(12):530. https://doi.org/10.3390/land9120530

EUROSTAT (2023) Sustainable development in the European Union—monitoring report on progress towards the SDGs in an EU context. Available online: https://ec.europa.eu/eurostat/documents/15234730/16817772/KS-04-23-184-EN-N.pdf/845a1782-998d-a767-b097-f22ebe93d422?version=2.0&t=1688373085450. Accessed 19 December 2024

Hensel M (2019) The rights to ground: integrating human and non-human perspectives in an inclusive approach to sustainability. Sustain Dev 27:245–251. https://doi.org/10.1002/sd.1883

Hensel M (2021) Integration of architecture and environment—the distinctive traits of embedded architectures. Archit J 12(637):15–19

Hensel M (2022) Embedded architectures: charting its traits en route to architecture and environment integration. In: Kanaani M (ed) Routledge companion to ecological design thinking: healthful Ecotopian visions for architecture and urbanism. Routledge, London, pp 460–470

Hensel M, Sunguroğlu Hensel D (2023) Architectures of the critical zone: architectures and environment integration en route to designing environments. In: Ulrich Hensel M, Sunguroğlu

Hensel D, Binder CR, Ludwig F (eds) Introduction to designing environments: paradigms & approaches. Springer, Cham, pp 183–207. https://doi.org/10.1007/978-3-031-34378-0_10

Hensel M, Sunguroğlu Hensel D (2024) Nature-integrated design—opportunities and challenges. In: Clements-Croome D (ed) Intelligent buildings. ICE Publishing, London, pp 95–106

Hensel M, Sunguroğlu Hensel D (2025) Re-grounding architecture(s). In: Leatherbarrow D, Shi Y (eds) Communicative construction. Artifice Press, London

Ingold T (2010) Footprints through the weather-world: walking, breathing, knowing. J R Anthropol Inst 16(s1):S121–S122. https://doi.org/10.1111/j.1467-9655.2010.01613.x

Lindgren E, Elmquist T (2017) Ecosystem services and human health. In: Oxford encyclopedia of environmental science. Oxford University Press, Oxford. https://doi.org/10.1093/acrefore/9780199389414.013.86

Nedd R, Light K, Owens M, James N, Johnson E, Anandhi A (2021) A synthesis of land use/land cover studies: definitions, classification systems, meta-studies, challenges and knowledge gaps on a global landscape. Land 10(9):994. https://doi.org/10.3390/land10090994

Our World in Data (2024) 'Land Use—How is Humanity using Earth's land? And how can we decrease our land use so that more land is left for wildlife? Available online: https://ourworldindata.org/land-use. Accessed 02 January 2025

Pastur GM, Peri PL, Lencinas MV, Garcia-Llorente M, Martin-Lopez B (2016) Spatial patterns of cultural ecosystem services provision in Southern Patagonia. Landsc Ecol 31:383–399. https://doi.org/10.1007/s10980-015-0254-9

Söderlund J (2019) The emergence of biophilic design. Springer, Cham

The World Commission on Environment and Development (1987) Our common future. Oxford University Press, Oxford. Available online: https://sustainabledevelopment.un.org/content/documents/5987our-common-future.pdf. Accessed 19 December 2024

United Nations Department of Economic and Social Affairs (2024) Sustainable development—the 17 goals. Available online: https://sdgs.un.org/goals. Accessed 19 December 2024

United Nations—Progress towards the Sustainable Development Goals (2024) Available online: https://unstats.un.org/sdgs/files/report/2024/SG-SDG-Progress-Report-2024-advanced-unedited-version.pdf. Accessed 19 December 2024

Weisser W, Hensel M, Barath S, Grobman YJ, Hauck TE, Joschinski J, Ludwig F, Mimet A, Perini K, Roccotiello E, Schloter M, Shwartz A, Sunguroğlu Hensel D, Vogler V (2022) Creating ecologically sound buildings by integrating ecology, architecture, and computational design. People Nature 5(1):4–20. https://doi.org/10.1002/pan3.10411

Chapter 3
Methodology, Strategies, and Actions for the Sustainable Development of Rural Inner Areas

Alessandra Battisti, Maria Valese, and Herbert Natta

Abstract This chapter investigates the potential of a systemic, multi-scalar, data-driven approach to support the definition of strategies and actions related to the development of rural inner areas. The proposed four-step methodological framework aims at managing the complexity of these marginal territories, to enhance the effectiveness of processes of regeneration. The preliminary collection and elaboration of spatial data, including demographic, economic, cultural, environmental input, enables an analysis entailing multi-scalar, multidisciplinary dimensions, integrating information on the value components of the natural, tangible, and intangible cultural heritage, along with the wealth of human and social capital available. This analytic phase brings to the fore the primary features of the identity of rural inner areas, ultimately leading to an operative, holistic vision for engaging local stakeholders in a participatory co-design process. The initial exploration is functional to a subsequent phase of synthesis that focuses on recognizing systemic interconnections, together with active and potential networks, with the aim to select neuralgic nodes that can maximize the impact of regeneration. In this way, the structural remoteness and marginality of rural inner areas is counterbalanced by co-creation with local stakeholders, and activation of physical and narrative interconnections between selected points of interests (POIs) that serve as new polarities (services, functions, infrastructures, etc.) that can enhance accessibility and attractiveness. This perspective opens the way to moving beyond the traditional urban/natural dichotomy, providing new classification models and initiatives of transformation to foster the sustainable growth of rural areas by reconnecting them to the social, cultural, and economic networks of their surrounding territories.

A. Battisti (✉) · M. Valese
Department of Planning, Design and Technology of Architecture (PDTA), Faculty of Architecture, Sapienza University of Rome, Rome, Italy
e-mail: alessandra.battisti@uniroma1.it; maria.valese@uniroma1.it

H. Natta
CNR-IMATI – Institute of Applied Mathematics and Information Technology, Pavia, Italy
e-mail: herbert.natta@ge.imati.cnr.it

A. Battisti et al. (eds.), *Urban-Rural Dialogue in Green City Design*, Designing Environments, https://doi.org/10.1007/978-3-031-88318-7_3

Keywords Rural areas · Sustainable development · Data-driven · Digital narrative · Co-design · Mapping

3.1 Introduction

Rural areas are a vital component of our global landscape, fostering agriculture, preserving natural ecosystems, and providing essential resources (ICOMOS 2017; Sypok and Marinchenko 2023). There is no single definition of rurality. The concept has progressively shifted from identifying a specific land use, traditionally presented as a dichotomy vis-à-vis the urban environment, to including the emergence of hybrid, *rurbain* contexts in which the rural components are integrated with urban needs (Antrop 2004; van Vliet et al. 2019).

To reconcile the different national terminologies and establish a comparative framework, the United Nations Statistics Division (UNSD) proposes a definition based on the degree of urbanization (UN Statistical Commission 2020), considering as 'rural' those territories where most of the population lives in low-density areas (less than 5000 inhabitants per km^2) (UN Statistical Commission 2020).

Furthermore, this demographic marginality is often interlinked with a 'inner' aspect tied to the level of remoteness. The definition of marginalization is not monolithic but depends on the context. In Europe, it is linked to a core-periphery model based on the distance from the nearest urban centre (more than 40 min by car) (NSIA 2013; Basile and Cavallo 2020). In Italy, the geography of inner areas is institutionally defined by the National Strategy for Internal Areas (SNAI), through a composite indicator based on the level of accessibility to primary services and infrastructures: mobility, health, and education. These fundamental rights are extrapolated as the distances from stations of the national railway network, secondary schools, and hospitals with emergency rooms, with the minimum threshold of accessibility set at a 20-min drive from the nearest hub. This results in a classification of inner areas into three categories of outlying territories: intermediate, peripheral, and ultra-peripheral areas, based primarily on distance expressed in terms of the time needed to reach the main service point of interest (Barca et al. 2014). In this sense, marginality is not only a spatial condition, but a process closely bound to socio-economic transformations (M'alikov'a et al. 2016).

Rural areas are directly involved in the dynamics of contemporary global urbanization. On one hand, there is the threat of such territories seeing their entire ecosystems endangered by the expansion of the built environment, as well as by depopulation trends and the rise of economic sectors less connected to agricultural production (Eurostat 2021). At the same time, these areas can play a strategic role in the sustainable development that is crucial to optimizing natural resources while also safeguarding cultural diversity (European Network for Rural Development 2020).

The United Nations' 2030 Agenda includes, among its Sustainable Development Goals (SDGs), ending hunger and achieving food security (SDG2). SDG2 target 2.a focuses attention on "increasing investment, including through enhanced international cooperation, in rural infrastructure, agricultural research and extension services, technology development", while prioritizing the sustainable development of rural areas. However, the World Social Report 2021 (United Nations 2021), entitled "Rethinking Rural Development", promotes a radical shift in the traditional rural-urban dichotomy, encouraging an "in-situ urbanization" process, under which the material standard of rural populations would be raised to urban levels. In fact, the WSR 2021 underlines the systemic endangerment of rural areas, which is where 60/70% of the population of low- and lower-middle-income countries live, as do approximately 80% of the people who find themselves below the poverty line. The report also states that many rural areas are witnessing severe depletion and degradation of natural resources, contributing to climate change and the recurrence of zoonotic diseases, such as COVID-19.

In the decades to come, depopulation will occur in European inner areas in Spain, France, Germany, Poland, Slovenia, Romania, the Czech Republic, Greece, and Italy (Wachter 2001), due primarily to two factors: emigration of the working-age population to big cities, plus the ongoing ageing of the population in the inner areas themselves, a trend generally accompanied by low birth rates. Though depopulation is normally viewed as a symptom rather than a cause of rural decline, whose chief determinants are low levels of accessibility and economic opportunity, it still constitutes one of the main factors to be addressed in managing the complex changes that are taking place in inner areas, leading to the emergence of new political visions which, at least in their declarations of intent, increasingly shift their focus from "agriculture" to "rural development", an objective to be achieved through policies of tourism and regeneration. In a far more incisive way than other forms of planning, that drawn up for inner areas tends to present itself as a generalized approach to achieving change in society (Slee 2020), and in the overall framework for living, by analysing points of diversity and proposing "custom-tailored sites" within the communities of residence (Dax and Fischer 2017).

The European Union considers rural development to be the second pillar of the European Common Agricultural Policy (CAP) supported by the European Agricultural Fund for Rural Development (EAFRD). EU countries implement EAFRD funding through Rural Development Programmes (RDPs), fostering knowledge transfer and innovation in agriculture while aiming to achieve a balanced territorial development of rural economies.

The EU has also defined a specific long-term vision for rural areas, entailing recognition of their social, cultural, and economic potential and upgrading of their inclusion and connectivity. By 2040, it aims to enhance the attractiveness of rural areas by providing public transportation and services while upgrading local digital infrastructure and levels of digital literacy. The fact is that rural areas account for 83% of the EU's territorial extension (European Commission 2021). Almost half of these areas qualify as remote rural zones, as most of their inhabitants live more than 45 min away by car from the nearest city (European Commission 2021). However, a large part of Europe's cultural and natural heritage is situated in these very same rural areas. Traditions and social practices tied to agriculture, farming and food-making are deeply engrained in almost all European cultures, contributing to a shared identity and sense of belonging.

On the one hand, the development of rural areas represents a top policy priority, given their quantitative and qualitative importance, though, at the same time, there is considerable risk involved in preserving the fragile equilibrium between anthropic and environmental dynamics at the core of the tangible and intangible rural heritage. The growth foreseen and encouraged in economic investment, together with the upgrading of services, infrastructures, and digitalization, is expected to generate new accessibility and visibility for remote rural areas, increasing their capacity to attract temporary external communities. Such scenarios call for the formulation of strategies for managing the resulting cultural, social, and economic impact. The EU Rural Action Plan (RAP) articulates and turns into action the above-mentioned long-term vision for rural areas, encouraging the creation of innovative, community-led ecosystems, to leverage the role of sustainable agriculture in efforts to diversify economic activities and attract innovative businesses.

The economic diversification of rural areas represents one of the main guidelines of international strategic plans, with significant potential to reduce poverty, placing the agricultural sector at the centre of rural-development strategies, to create better jobs and, at the same time, enable the growth of non-farming activities within the rural economy (International Labour Organization 2019).

Reviews of various European strategies show that the operating instruments used in France for inner areas take the form of "Contracts of Reciprocity", under which cities and their surrounding countryside areas agree to improve the sustainability of rural services, economic development, and the environment. A different approach is followed under Belgium's "Service Design" strategy for Western Flanders, which focuses more closely on improving transportation and mobility in towns, while involving citizens in decision-making processes (ENRD 2018).

In Sweden, emphasis is placed on increasing services in inner and rural areas by supporting initiatives designed to heighten accessibility to those same services, both public and private. Spanish policies for sustainable rural development are enacted through an integrated approach that supports rural infrastructures, utilizing and implementing ICT, together with a wide range of rural services, with initiatives including incentives for economic development to counter depopulation. In Italy, an integrated national strategy of socio-economic development has been formulated in recent years, involving approximately 13 million citizens, and including four specific pilot projects geared towards supporting innovation in local services (Garrod et al. 2006).

A second point deserving attention concerns the social costs generated by processes of production and consumption, pointing to the need to acknowledge the existence of dynamic relations between the use of natural resources, environmental quality, health and wellness, while moving away from compartmentalized approaches based solely on risk evaluation, in favour of more integrated, ecosystemic perspectives that focus, in terms of formulating meaningful policies, on prevention, heightening of awareness and overall well-being (Shane 2005).

In this context, hydrogeological imbalances, inadequate maintenance, loss of biodiversity, deterioration of the countryside on a human scale and loss of traditional know-how appear to be top priorities for intervention.

The third vitally important consideration is the social disadvantages tied to shortages of basic services: education, healthcare, mobility, and connectivity. Such shortcomings are traceable, in part, to the low population density of these territories, which makes it difficult to organize efficient production and supply of services (Smith and Ripp 2020).

What is needed is a scenario that can interpret an evolving territorial system by addressing topics of space, economics, and social and environmental concerns, tracing their future paths of development, to identify prospective initiatives, which should be judged in terms of:

- safeguarding of the territory and local communities.
- enhancement of natural and cultural resources, plus sustainable tourism.
- systems of agri-food production and local development.
- energy savings and local renewable-energy production facilities.
- crafts know-how and artisanal skills.

This perspective is endorsed by many authors (Shane 2005; Smith and Ripp 2020; Garrod et al. 2006) and is used to assess the various projects being implemented under national and regional strategies promoting the valorisation of population centres in inner areas, as well as the revival of rural services and the quality of life, to which end, systemic strategic guidelines can be identified on multiple scales, both horizontal and vertical, in the interests of:

1. designing a model of regional governance with operating objectives and long-term resources, commitment, and leadership, active on various scales, from the local to the regional, as well as through inter-municipal cooperation, with contributions from both public and private stakeholders.
2. enhancing minor historic population centres and meaningful distinguishing features of historic rural landscapes through a process that reinterprets as potential the complexity, and the multiple manifestations, of the value of inner areas with respect to considerations of history, identity, and rural life.
3. improving relations between historic settlements (historic towns, burghs, villages) and their surrounding territories by establishing principles and criteria with a view to upgrading and guaranteeing the wellbeing of the landscape and the territorial environment.
4. developing and confirming a functional role for the historic resources of minor population centres through the dissemination of good practices for the transformation, as well as the preservation, of rural architecture and the landscape, with positive repercussions for the environment as well.
5. fostering economic diversification, to promote local resilience, with a focus on the endogenous management of resources and, whenever possible, initiatives designed to encourage businesses to establish operations in rural inner areas.

6. promoting active engagement of established communities in determining the actions and initiatives of public authorities whose decisions have a direct or indirect effect on issues affecting the landscape in the territories they govern, as regards regional, provincial, municipal, and sector-based planning.
7. transforming, preserving, and enhancing the historic landscape while reinforcing synergies between policies addressing the environment, the countryside, urban planning, architecture, culture, and rural areas, with respect to topics of resilience and the defence of the landscape and the environment.
8. strategic, multi-settlement planning through identification of accessible services shared by different locales, making possible the establishment of multi-functional service hubs, to optimize resources while maximizing the impact of ICTs for the public.
9. creating broadband infrastructures and disseminating digital educational opportunities through courses and workshops for both private citizens and business operators, plus the organization of pilot projects.
10. promoting and creating a monitoring system through the application of indicators capable of controlling the territorial transformations of inner areas, minor historic population centres and rural landscapes, as well as the effectiveness of the initiatives proposed.

This process calls for the development of comprehensive, inclusive planning strategies to be adopted and implemented under a multi-action, multi-stakeholder participatory approach.

3.2 Methodology

The key action for the renovation of the small-scale realities in the villages of rural European inner areas must be found in the implementation of local activities and the encouragement of new modes for activating processes of circular economics capable of attracting people and investments without compromising the delicate equilibrium of the local ecosystem. Steps must be taken to identify and develop new scenarios of innovation, to support communities in inner settlements while promoting the competitiveness of small- and medium-size companies; a vocation for territorial innovation can be augmented through material and immaterial interconnectivity, linking engagement in socio-economic development with the power of attraction of the local area. Accessibility and safety must also be ensured, using environmental and ecological factors as leverage, to reach full integration between places and infrastructures, as part of a developing model in which the users (communities) can be "prosumers". This model needs to be as interactive and integrated as possible, to identify new perspectives of operability within a well-rounded approach.

To this end, two Italian case studies have been selected: Val Gandino and Piaggine. Both are inner areas that present the common themes of scarce competitiveness of their commercial sectors, plus a lack of interconnectivity. These territories need to:

- increase awareness of the potential of their territorial areas and of the capabilities of their SMEs.
- diversify the types of services and products offered.
- specialize in innovative, attractive activities, along with infrastructures tailor-made for the end users, in terms of both services and eco-friendly reception facilities.
- enhance local resources related to green circular-economic activities by increasing the authenticity of the historical cultural heritage and by upgrading the safeguarding, management, and assurance of sustainable access to environmental resources.

Overall, the proposed methodology is structured in four phases:

1. Territorial framework identification: this phase involves the collection of data and mapping procedures, as well as identification of the users' needs and demands. It consists of desk research activities resulting in performance of a cognitive survey, plus careful study of the areas involved through structured mapping and cataloguing of anthropic, biophysical, energy-related, social and cultural data, all designed to develop a cost-benefit analysis (preliminary feasibility assessment), complete with identification of the environmental, economic and social benefits expected by the community, formulation of the legal framework, and identification of the clusters of actors to be involved. GIS applications can be used to collect and map data on:

   a. the socio-economic and demographic context, in order to identify the distinctive characteristics of the community by studying data from sources such as statistical institutes, local government databases, the distribution of questionnaires, and the identification of initial macro-areas of interest.
   b. environmental, facility and mobility macro-systems, in the interests of obtaining an integrated view and understanding of the peculiarities of the context in which the initiative is to be launched.
   c. active realities operating in the area, such as associations, social or cultural cooperatives, local citizens networks, economic and social enterprises with a special focus on local SMEs (farmhouses, B&Bs, diffuse hotels in historic villages).
   d. energy needs and consumption of relevant users, to draw up an energy model based on renewable energy sources and the development of models of collaborative economics.
   c. public and private building stock in the case study, to quantify the areas suited to implementation of the process of regeneration.

2. Stakeholder engagement: following the desk analysis and the clustering process of the first phase, this second phase shifts the focus to formalising the engagement of local communities, taking into consideration both top-down and bottom-

up approaches. The emphasis is on qualitative aspects and parameters collected through in-depth analytical activities, such as focus groups, roundtables, and other occasions of participatory planning. Focus groups will be organized according to the needs and features of each case study. The engagement of different categories of stakeholders makes it possible to collect points of view that can prove valuable in structuring directly accountable local governance, based on a set of principles, rules and procedures shared among the participants in the different focus groups.

3. Co-design/Co-creation: in this phase, the focus shifts to operational research carried out through co-design and co-creation initiatives. It will be custom-tailored for each case study, based on the key features to have emerged during phases 1 and 2. Transformation scenarios will be developed, in accordance with the specific characteristics of the inner area, by:

    - augmenting the territorial resources for developing a retooling plan that targets different types of potential new residents, temporary or permanent (e.g., singles, families, old people).
    - configuring the type of involvement to match the territory in question: experiential, social, naturalistic / sporting, cultural, historical, art tourism.
    - promoting innovative forms of hospitality, such as diffuse hotels, networks of bike hotels, park hospitality services (e.g., zero-impact accommodations), and pathways of ecological brands.
    - enhancing educational factors as promoters of environmental themes.
    - reconsidering the seasonal nature of tourist packages, to ensure flows throughout the year (e.g., promotion of events tied to agricultural activities, such as grape, olive, or chestnut harvests, plus combinations of landscape / cultural routes).
    - promoting a model of integrated paths and events to support tourism in the local territory.

A process of cross-fertilization could facilitate achievement of a convergence of territorial development policies that increases the advantages of the alignment of the respective socio-economic systems. In this way, a virtuous process leading to revitalization of the territory could be triggered, to implement an efficient and functional governance mode under which all the stakeholders of the territories in question are actively represented and involved.

4. Development of innovative narratives and storytelling: the (multilevel) narrative dimension of the territory must become a driver of new forms of sustainable development. Sustainability and eco-compatible innovation offer the opportunity to enhance the attractiveness of rural inner areas, both for new residents and for tourist flows, while considering the delicate aspect of mitigating the related environmental impact. A community heritage enriched by natural capital, culture and local craftsmanship that is little known outside the local territory, and often penalized by poor accessibility and lack of material and immaterial infrastructures, represents a wealth of potential power of attraction.

3.3 Case Studies

3.3.1 Rural Inner Areas as Experimental Phygital Labs

The methodology has been applied to the formulation of regenerative strategies for two selected case studies: the Gandino Valley (Lombardy, Italy), and the village of Piaggine (Campania, Italy). They are both apt examples of Italian rural inner areas threatened by a depopulation trend that risks exacerbating the negative effects of their physical isolation. However, they are in two different geographical contexts: the Gandino Valley is historically connected to the productive landscape of the industrial north, while Piaggine is found in the central-most portion of one of Italy's largest national parks.

They also play different roles in the rural make-up of the local ecosystem. In the case of Gandino, while most of the land cover is natural, only a small portion is farmed, and the history (and current state) of its usage leans more heavily towards industrial activity. Piaggine's development as a settlement, on the other hand, was closely tied to the pastoral production chain, which still plays a noteworthy role, turning out high-quality food products.

In both cases, the proposed initiatives were aimed at activating local ecosystems, identifying resources, engaging stakeholders, and providing, through participatory activities, innovative narratives to recast the rural condition as a driver of development. Furthermore, the differences in the relations (past and present) of the two areas selected with their nearest urban polarities render them meaningful examples for reformulating rural development less in terms of the rural-urban dichotomy than from the perspective of a *rurbain* interconnection and hybridization (Haller and Branca 2022).

3.3.2 Val Gandino—The Invention of a Metropolitan Mountain System

The Val Gandino (Gandino Valley) is a mountain area located in the Alps, in Northern Italy, 20 km from the City of Bergamo. This relative proximity to an urban polarity has generated, over time, a specific territorial system. Proto-industrial economic activity, connected primarily to textile production, dates to the thirteenth century, and still constitutes the predominant economic sector (by number of employees) (ASIA 2022). However, land-cover figures point to a sizeable prevalence of woodland (35% of the total surface area, source: CLC 2018), and high-mountain pastures (37%), with only a small portion (10%) devoted to agriculture, including cultivation of a specific type of corn (Fig. 3.1).

From an administrative perspective, the Gandino Valley is a system of five municipalities (Gandino, Cazzano, Casnigo, Leffe and Peja). In line with the demographic trend for European rural areas, it has suffered a progressive depopulation

(−5% since 2015) (ISTAT 2022), along with increase ageing (average age 45–50) (ISTAT 2022). Younger residents prefer to move closer to Bergamo, or to neighbouring municipalities offering better connections.

Public transportation service is limited to two bus lines which connect, at a low frequency, the valley's five settlements. This, in effect, expands the physical distance from Bergamo by nearly 2 h of actual bus travel. As a result, the motorization rate in the valley is far higher than in the urban area.

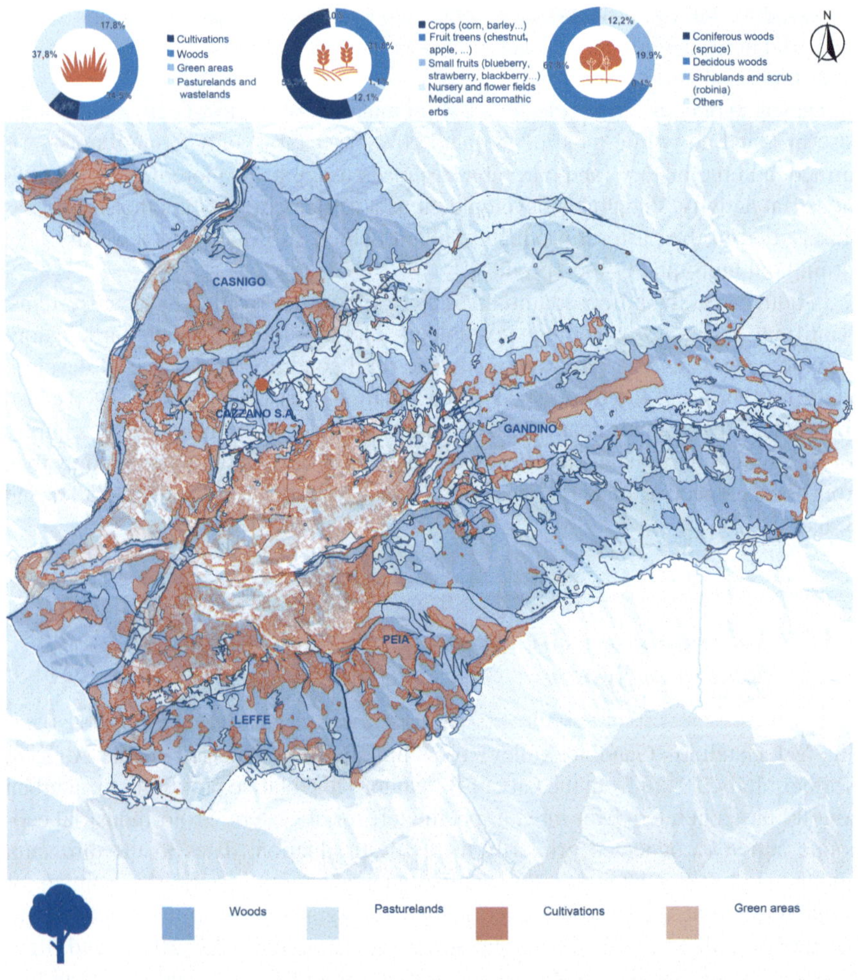

Fig. 3.1 The image shows the composition of green areas in the Gandino Valley. It reveals the quantity and quality of agricultural cultivation, which accounts for less than 10% of the total surface area, which is planted predominantly with corn and fruit trees. (Data sources: Territorial Coordination Plan of the Province of Bergamo (PTCP, 2022), Italian Alpine Club, Authors: Daniele Avenoso, Federica Pezzoli)

As determined under the methodology, the approach to the study of the Gandino Valley began with a preliminary phase of data collection designed to identify and analyse the chief characteristics of the territory. In addition to the above-mentioned demographic, economic and land-cover data, specific attention was devoted to analysing, at different scales, the system of infrastructures and services, to highlight the valley's internal and external connections.

While vehicular mobility is limited by a lack of roadway infrastructures, a capillary system of pathways reconnecting the five settlements can be observed, pointing to possibilities for movement through the natural environment (Fig. 3.2). Furthermore, the Bergamo Provincial Mobility Plan includes, among its initiatives, reconnecting the five municipalities with a cycleway that, running along the Serio River, would pass through both the natural and the built environments.

The mapping of the services active in the area revealed the presence of elementary and secondary schools, but no establishments of higher education, and, overall, a scarcity of hospitals and healthcare centres, with the nearest found at a distance reachable by car in 15–30 min.

In preparation for the second phase, a preliminary stakeholder analysis identified the existing local networks and the actors who could potentially be involved in the regeneration process. The main finding was the presence of a diffuse commercial district called "The Five Lands of Gandino Valley": a consortium of 198 shops and commercial activities in the five municipalities, established to promote urban renewal, tourist events and territorial marketing.

The connection with this composite subject has opened a path to direct engagement of local government bodies and civic associations through the organization of roundtables and focus groups to discuss the results of the data analysis and to summarise its outcomes in a set of core principles to be followed in formulating strategic guidelines during the third phase of the workflow.

Based on this premise, a co-creation process leading to an inter-municipal, joint public-private subject was proposed, with the aim of activating a multi-scalar regeneration process based on the idea that a network of municipalities in rural areas could work as an interconnected metropolitan network (Fig. 3.3).

The core of the co-designed proposal consisted of expanding the partnership into a multi-layered system of systems, under which the development of renewal and retooling initiatives for the built environment would be supported by a plan for enhancing the area's accessibility, while also promoting a digital and ecological transition centred around local agricultural production.

As a first step, it was proposed that new bus services be activated, to provide an intermodal connection between the five municipalities and the planned extension of a direct tram line linking Bergamo with Albino (a small town located 8 km from the Gandino Valley). Furthermore, the existing pathways were to be reconnected to form a unified walking and cycling system, promoting sustainable mobility between the settlements (Fig. 3.4).

This strategic plan supports the retooling and reactivation of some significant elements of the built environment, with the insertion of a new health centre with distributed access points, the renovation of existing, but not currently used,

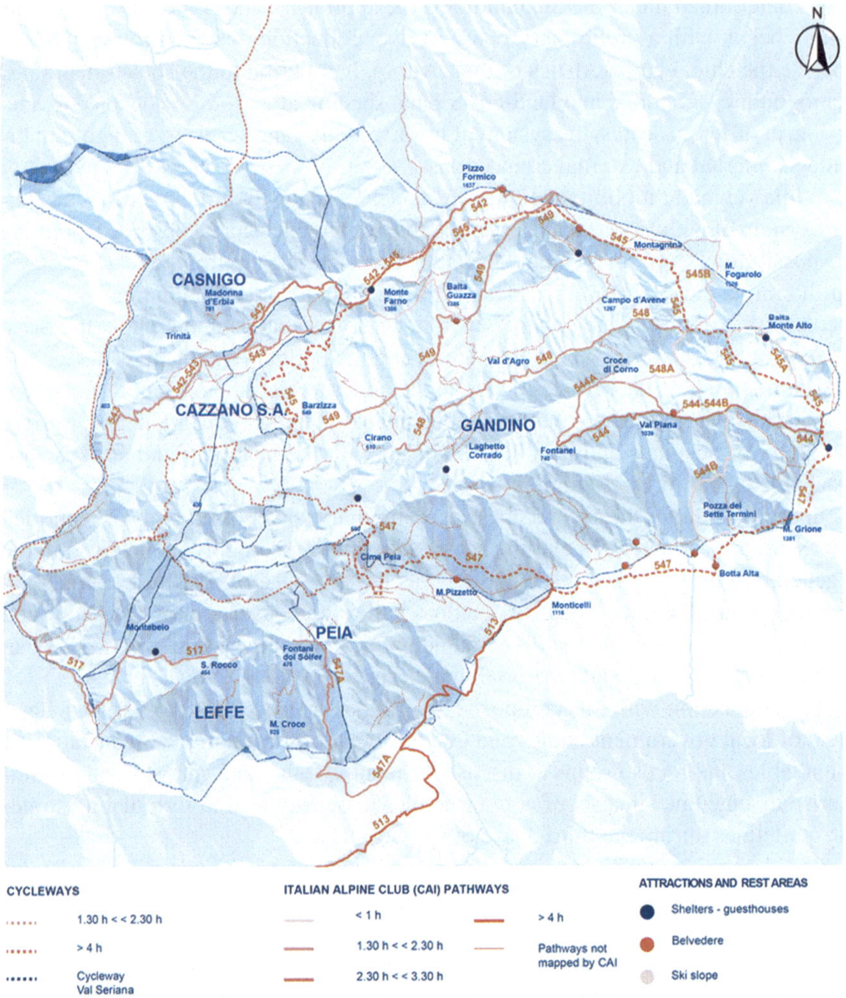

Fig. 3.2 This map shows the existing network of pathways and cycleways in the valley, together with the figures for ascending and descending altitudes (Source: Italian Alpine Club, Authors: Daniele Avenoso, Federica Pezzoli)

high-mountain shelters, and the creation of a multi-building diffuse hotel to encourage sustainable tourist traffic.

Moving on to the second and last phase of the methodology, a new network of POIs and pathways has been developed along six thematic itineraries (nature, culture, food, textiles, sports, history) (Fig. 3.5), in order to establish a common framework for communicating and promoting the overall vision of the strategic guidelines while enhancing the attractiveness of the area, both for new residents and to encourage sustainable tourist flows.

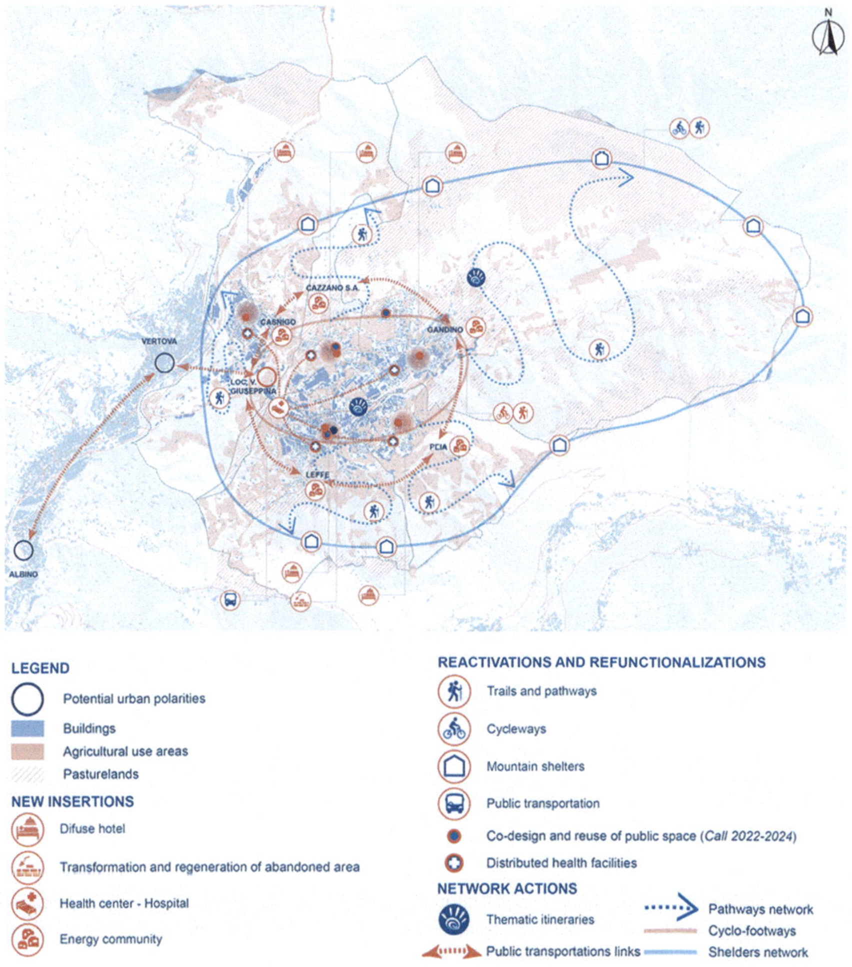

Fig. 3.3 The synthesis of the regeneration master plan presents the main strategies on a territorial scale: (i) the reconnection of pathways and cycleways as interlinked itineraries; (ii) the promotion of high-mountain exploration through thematic paths and the reactivation of mountain shelters; (iii) the creation of a diffuse hotel; (iv) the activation of an energy community; (v) the activation of a health centre; (vi) the activation of new public transportation lines (Authors: Daniele Avenoso, Federica Pezzoli)

In fact, this narrative level has contributed to increasing movement and exploration throughout the valley, identifying new destinations where history and traditions can be discovered, along with exceptional natural and cultural features, at the same time as initiatives of cultural exchange highlight crafts know-how and agricultural products of excellence. The activation of thematic narratives in the form of physical pathways through the valley has been supported with GPS applications that make it possible to deliver place-specific digital contents (with natural sounds, local events,

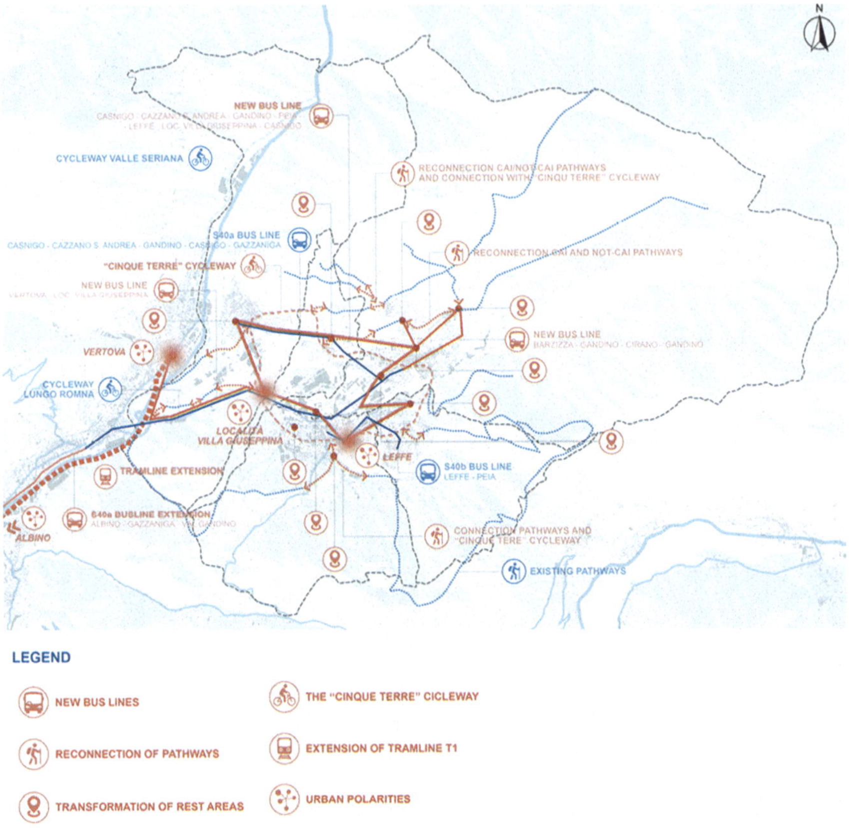

Fig. 3.4 This map shows the proposed strategic mobility plan, based on the inter-modality between the network of public transportation, the cycleways, and the reconnection of the mountain footways. (Authors: Daniele Avenoso, Federica Pezzoli)

recollections of the past, traditional songs etc.), generating low-impact phygital experiences.

In this way, the enhancement of physical accessibility of the valley, and the empowerment of its services, has been interlinked with a renovated, multi-layered territorial image constructed on a foundation of local cultural and natural core elements and shaped by a contemporary narrative format (multimedia digital contents). This process has resulted in the reinvention of the Gandino Valley as a metropolitan mountain system, making possible the development of contemporary lifestyles in equilibrium with the specific features of a predominantly natural environment.

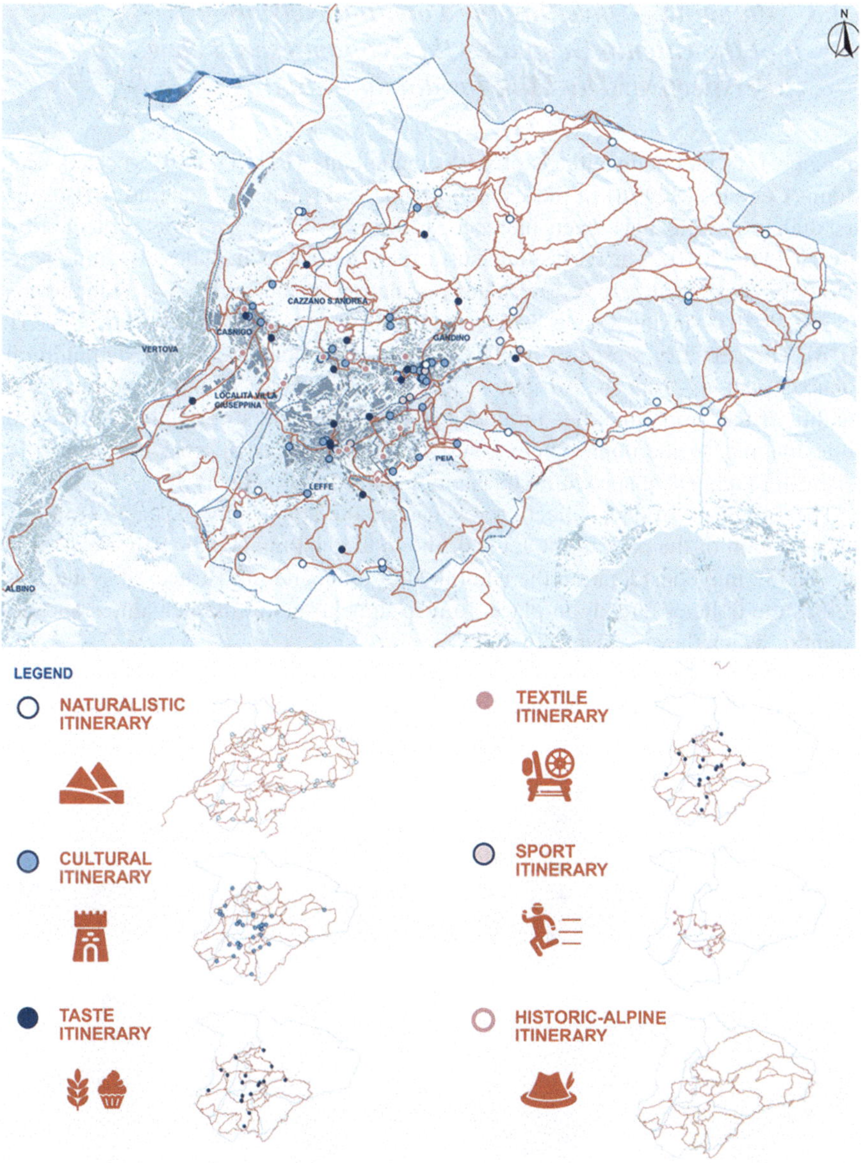

Fig. 3.5 The six thematic itineraries function as a system for reconnecting the valley's POIs: they define preferred exploration pathways, along with the possibility of linking temporary activities, functions, or levels of intangible information to the exploration of the nodes selected. (Authors: Daniele Avenoso, Federica Pezzoli)

3.3.3 *Piaggine—Unveiling the Potential Network of the Cilento Geopark. Old Economics in a New Perspective: The Milk Production Chain*

Piaggine is a mountain village of 1300 inhabitants (ISTAT 2018), located near Mount Cervati (1899 m) in the Cilento Geopark (Province of Salerno, Italy), the first Italian park to have been included in the network of geoparks protected by UNESCO (in 2010). The park's total surface of 181,048 ha includes 80 municipalities, only six of which have a resident population of more than 5000 inhabitants (ISTAT 2018), making for an average population density of 100 in./km^2 (ISTAT 2018). This territory presents a wide variety of geographical, natural, and cultural contexts, ranging from the Adriatic seacoast to the inland Apennine mountains: a condition that, over time, has produced overlapping layers of administration and policymaking, to the point where a systemic, multi-scalar approach is needed to establish Piaggine's place within this complex territorial network.

The first phase of data collection was carried out at a variety of scales, from the macro-region of the park to the level of the individual municipal territories, while also taking into consideration the natural, anthropic, and infrastructural system, as well as the policies already in place, to map the clusters of municipalities and the extent to which they overlap (Fig. 3.6).

The analysis detected similarities involving natural, cultural, and environmental conditions; for example: the presence of the same river, location in the same

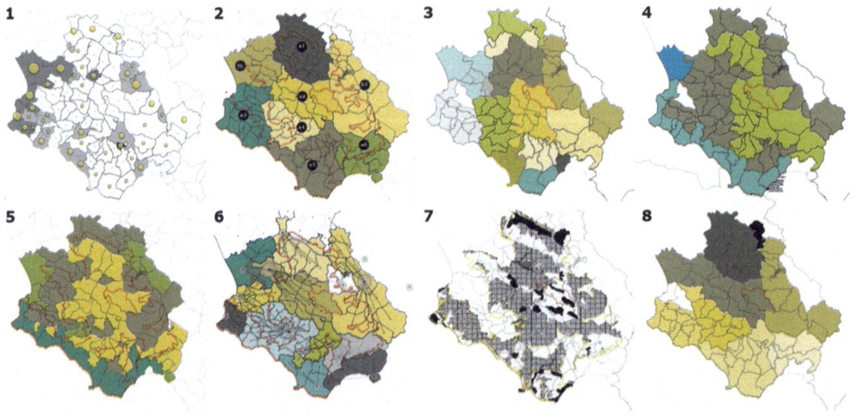

Fig. 3.6 The cartographic diagrams visualize the data collection and elaboration at the scale of Cilento Geopark, taking into consideration: (1) the distribution of population density among the municipalities; (2) the territorial development systems defined under the Regional Development Plan; (3) the existing mountain communities; (4) the altimetric zones; (5) the economic regions; (6) the Labour Market Areas (LMAs); (7) the policies for safeguarding the natural environment; (8) the landscape structure. (Sources: processing of data from: ISTAT, resident municipal population by age, sex, and marital status, 2018; ISTAT, Labour Market Areas, 2014; Regional Development Plan of Campania, 2008. Author: Maria Valese)

altimetric zone, involvement in the same local action group (GAL) or mountain community, etc. This preliminary analysis characterised Piaggine as an inner mountain, low-population-density area with protected sites of agroforestry, natural and cultural interest, plus a connection, in terms of attraction of the labour force, to the municipality of Roccadaspide. Furthermore, Piaggine is part of the economic region of the Cilento hinterland (or "Heart of the Park"), characterized by trends of ageing and depopulation, plus a generalized lack of services.

During the first phase, specific attention was paid to analysing the networks activated by the economic production chains. Paradoxically, it was found that agro-pastoral activities, though deeply rooted in the area's productive history and traditions, and even though they continue to provide high-quality resources and excellent agricultural products, constitute an undervalued sector, in terms of both perceived quality and indexes of quantity (number of employees, number of companies etc.).

As a preliminary overview, on a territorial scale, two integrated networks of municipalities have been recognised: the first based on the presence and characteristics of the natural environment and resources, the second defined by the economic zone and its related connections. Piaggine stands at the intersection between the territory through which the Calore River flows (from the inner mountain area to the sea) and the area that surrounds the peaks of Mount Cervati and Mount Gebison. The river serves as both a natural and an administrative link to the coastal area, reconnecting the pair of economic regions (the Cilento hinterland and the semi-industrialized countryside) and the two altimetric zones (mountains and inner hills), a territorial relationship confirmed and further expanded by local networks (such as the Cilento Regeneration GAL).

The idea of enhancing this strategic connection by identifying and activating, on the one hand, a local circuit of similar municipalities, while, at the same time, establishing a territorial link between them and the surrounding areas potentially involved in circular economic chains, has been supported by a more detailed data collection and analysis of local agro-pastoral products of excellence, as well as speciality PDO (protected denomination of origin) foods.

It was found that the Piaggine territory is home to a specific race of cow, the *Podolica*, characterised by an elevated capacity to adapt to hostile environments, plus an ability to consume alimentary resources that have no other use, all while producing high-quality meat and milk.

The *Podolica* is farmed primarily in the Apulia Region (southeast Italy), where it is part of a certified Slow Food presidium, whereas in Piaggine this breed of livestock, along with its derivative products, are undervalued in terms of their potential contribution to sustainable development.

An analysis of local resources, together with the economic supply chains that link the agricultural and pastoral products to their processes of transformation, set the stage for preparation of the second phase of stakeholder engagement, with an assessment of the levels of synergy between local producers and the local government bodies of the territories involved in the process viewed as a starting point for achieving better conditions through a sharing of resources and human capital.

The project of a "sector" of municipalities and thematic itineraries designed to operate at different scales is tied to the organization of a variety of specific tourist offerings tailored to various time periods and target markets, in support of the underlying logic of a system of interconnected villages that makes possible exchanges and working relationships meant to optimise the existing supply chains.

It is under this approach that "La Via del Latte", or "Milk Route", was developed (Fig. 3.7), establishing a thematic itinerary between the municipalities, with the aim, on the one hand, of attracting tourists to Piaggine by promoting the rural setting, the local wine and cuisine, including traditional dairy products, all part of a modern reinterpretation of the pastoral environment, while, at the same time, creating economic synergies between rural tourism, programs of research and higher education, plus the marketing of local products, and subsidies for local farmers.

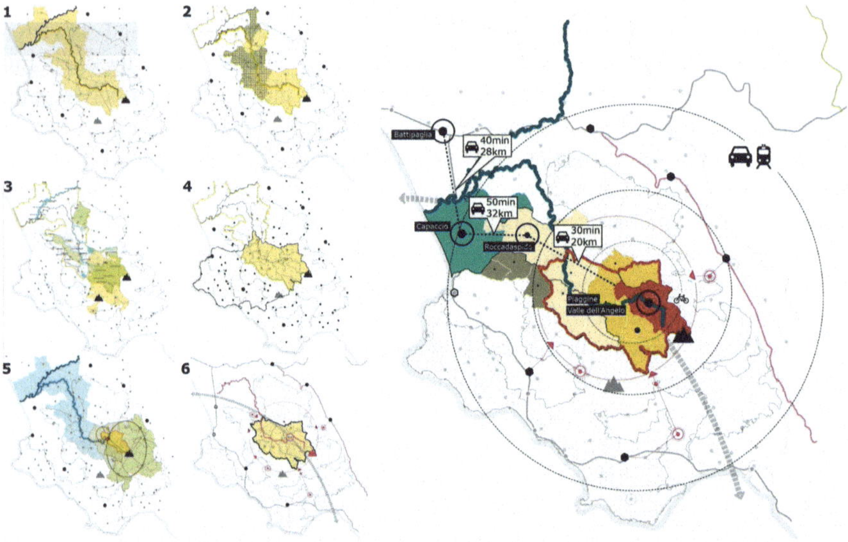

Fig. 3.7 The "Milk Route" ties together the main territorial networks identified as: (1) the municipalities through which the Calore River flows; (2) the areas of intersection between the economic regions (the Cilento hinterland and the semi-industrialised countryside), and the altimetric zones (hinterland hills and mountains); (3) the overlapping of the protected mountain areas between Mount Gebison and Mount Cervati; (4) the Cilento Regeneration GAL. The reconnection of these networks has given rise to two integrated visions: the first (5) based on the intersection between the river and mountain areas (centred around Piaggine); the second (6) resulting from the enhancement of the connections between Piaggine, the inter-municipal network and the nearby infrastructural polarities. The "Milk Route" combines the two visions into a single multi-layered concept. The itinerary of milk production and transformation establishes a connection between the various rural systems: the Sele plain, the hills of the Cilento hinterland, and the system of the Alburni and Cervati mountains. On the one hand, it proposes upgrading the territory through knowledge of local production techniques and tastings of typical dairy products, while, at the same time, economic development is promoted through a system of vocational training and thematic research centres supporting the continued development of the livestock sector. (Author: Maria Valese)

The main goal was to support the activation of a local network, at both the municipal and inter-municipal levels, centred around the economic chains generated by the products of agro-pastoral excellence identified. The *Podolica* cow, by giving rise to an entire rural system, made it possible to reconnect Piaggine with the municipalities of Valle dell'Angelo, Roccadaspide, Capaccio and Battipaglia, all in accordance with the concept of the Milk Route and the broader aim of integrating the dairy offerings of the Campania region.

The better-known, and more firmly established, agro-pastoral production activities tied to buffalo farming, involving the major population centre of Battipaglia and the countryside around the Sele River, should work as a driver for the promotion of the bovine economic chain of the Cilento inner area.

In this way, the stakeholders of the eight municipalities included in the inter-municipal circuit identified in the first phase have become directly involved in collaborative activities (roundtables, workshops etc.), moving from a preliminary data analysis of the circuit (Fig. 3.8) to the co-creation of development strategies.

Of particular importance to the development of a specific transformation scenario for the municipal territory, thanks to the interlink with the territorial and inter-municipal polarities identified (Fig. 3.9), has been the direct involvement of the municipal administration of Piaggine as the main stakeholder in the co-creation phase.

Based on this premise, the reactivation of the village entailed several measures involving mobility, restoration of the constructed heritage, the organization of the local area and the definition of relations between the village and the relevant itineraries, on the local and territorial levels, always in accordance with the numerous restrictions traceable to the Geopark regulations and to the variety of different administrative structures involved.

The management of the constructed heritage of the historic village is closely tied to these efforts to enhance and preserve the landscape: the project combines the reactivation of rural areas with the organization of physical space in the town.

The municipal territory of Piaggine is far bigger than the settlement per se, meaning that, to reconnect its main POIs (both cultural and natural), an entire network of pathways must be crossed.

In continuation of, and compliance with, the definition of "The Milk Route", the territory has been designed as a mountain park, providing various scenarios for rural tourist experiences tied to the farming of the *Podolica* cows. The cows are generally free to graze in the mountain area, in which outfitted rest areas, shelters and bridal paths have been established for sustainable exploration of the natural site. Closer to the town centre, there are plans for installing information points and spaces set aside for the sale of local products.

Furthermore, a new research centre for agro-alimentary studies has been proposed for the historic core of Piaggine, designed, in symbolic fashion, to be housed in the venerable Palazzo Vairo building. The aim of the initiative is to activate new functions tied to food education and the promotion of local products, providing spaces for a thematic library, a small store, a restaurant and some laboratories for advanced studies and research.

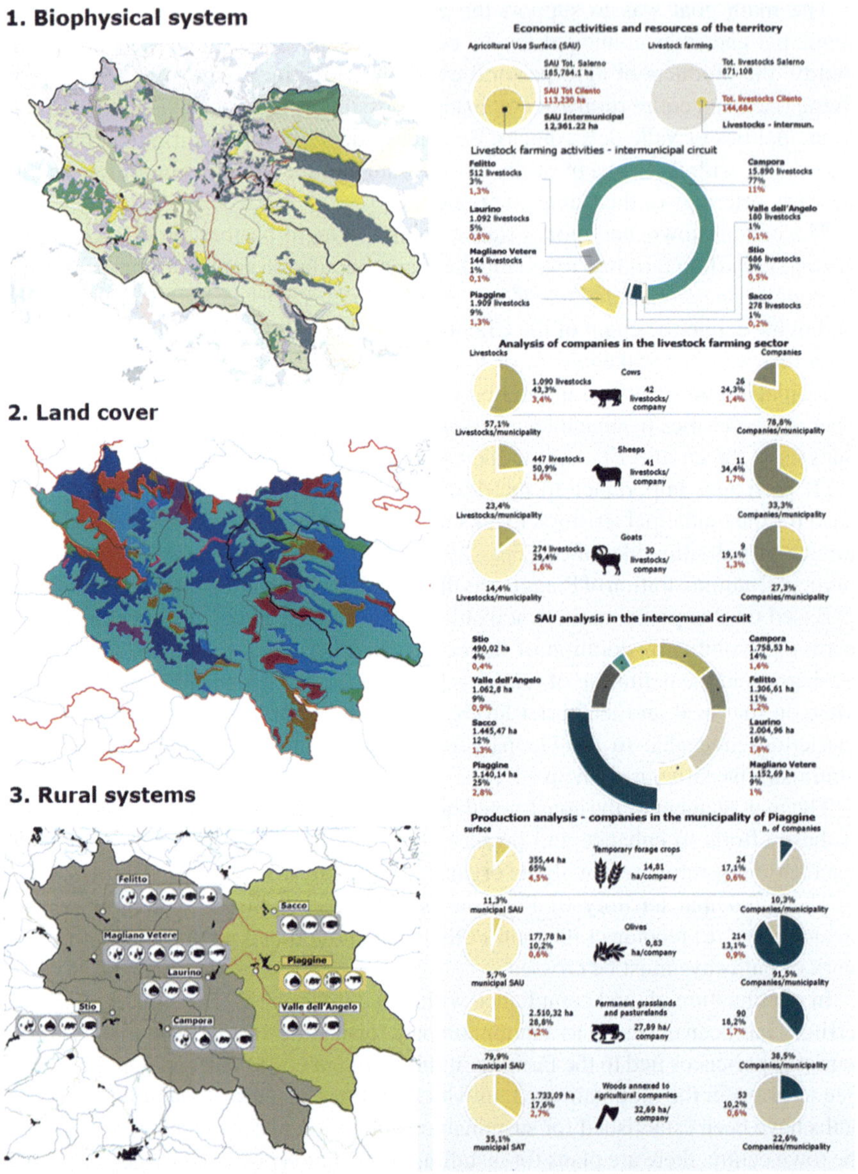

Fig. 3.8 The inter-municipal circuit, identified during the preliminary data-collection phase, has been analysed in greater detail, with particular attention given to the territorial systems and data sources tied to agro-pastoral production chains. The illustration shows the biophysical system (1), where the dominant elements of the land cover (2) are agricultural fields (in pink) and woods (in green), and more specifically: deciduous forests (23% of the total surface area) and open grasslands (temporary pastureland, abandoned land etc., accounting for 32% of the total surface area), plus the rural systems, which are formally divided in two (the Cilento hinterland hills and Mount Cervati), though both provide similar products of agro-pastoral excellence (cheeses, meats, wines, olive oils etc.). (Sources: processing of data from CLC 2018, ISTAT, Agricultural Census, 2010; Regional Agriculture Department of Campania, The Rural Territory of Campania, 2013. Author: Maria Valese)

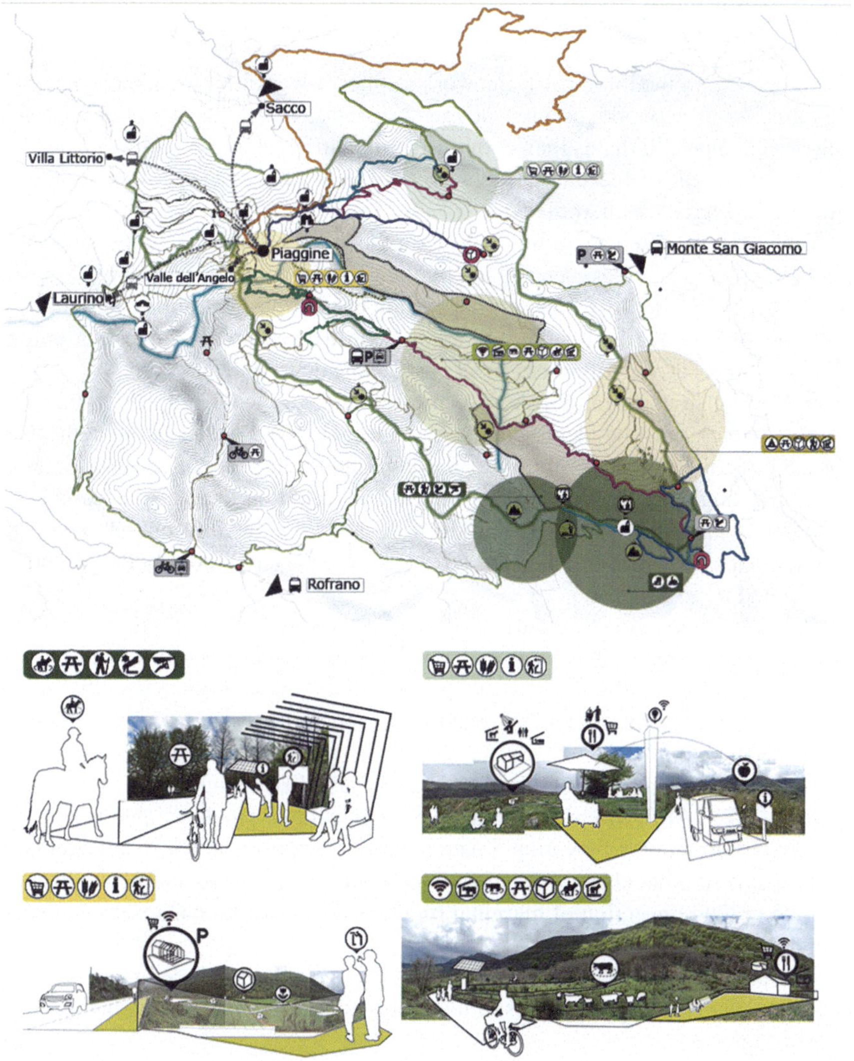

Fig. 3.9 The illustration shows the general plan of the mountain park in the municipal territory of Piaggine (the thematic itineraries and the POIs), along with samples of transformation scenarios, based on the concept of adopting low-impact technologies for operating in the natural environment, so as to minimise soil consumption while generating opportunities to produce immersive rural experiences through digital infrastructures, services and new functions also able to attract outside visitors. (Author: Maria Valese)

In this way, the concept of "The Milk Route", generated by a systematic data analysis, has evolved into a new narrative for the territory of Piaggine: the starting (or arrival) point of an itinerary enabled by a modern interpretation of traditional rural practices as a catalyst for development.

3.4 Conclusion

The proposed methodological framework defines a workflow for identifying and activating local networks as innovative ecosystems, based on an initial analytic, data-driven approach that is then combined with participatory and co-design activities, all in pursuit of the ultimate objective of developing regeneration strategies supported by territorial narratives.

Application of the approach to selected case studies reveals how the structural weaknesses of rural areas, determined by their marginal status, can be reimagined as a multi-layered, multi-scalar *rurbain* system of multiple centralities, enriched by their valuable cultural and natural features, with the resources in question potentially serving as drivers of processes of sustainable development.

In defining strategic initiatives, the core elements of the territory (both tangible and intangible), must be identified and reconnected at various levels: in a physical sense, to enhance the accessibility of the POIs selected; from an economic standpoint, to engage stakeholders, in the interests of including local resources in contemporary, innovative chains of production; and, on the information level, to promote narratives that enhance the territory's capacity to attract both temporary visitors and new residents.

The countermeasures to the demographic, economic and environmental endangerment of rural hinterland areas call for the activation of polarities of attraction able to overcome the disadvantage of physical remoteness through specific initiatives designed to reconnect and systematize the territorial resources while affording them new visibility.

In this sense, the use of digital technologies (like Geographic Information Systems, GPS, AR/VR, etc.) as a support for integrated pipelines able to convert data into information by creating a data infrastructure from which to generate different forms of contents and visualizations (maps, audio-video, etc.) has played a key role in the reinvention of marginal rural systems as relational spaces endowed with distributed centralities.

This premise encourages the integration of ever more advanced, sophisticated technologies in the various phases of the proposed methodology, to maintain a low-impact approach.

This became possible via the ongoing evolution and widespread use of artificial-intelligence models has led to the possibility, in the first phase of processing heterogeneous data sources by means of machine learning, so as to provide the input needed to simulate transformation scenarios in support of co-creation activities and the decision-making process. The integration of these tools with gamification methods can facilitate the engagement of the stakeholders and the local community, as well as the participatory development of co-designed transformation solutions.

What is more, gamification and digital technologies make it possible to design and create dynamic scenario-based experiences, which, placed in an overlapping dialogue with the physical site, generate new opportunities for providing immersive experiences of rural or cultural tourism.

The United Nation World Tourism Organization (UNWTO) has recognized the elevated potential of rural tourism for stimulating local economic growth and social change, given how it complements other economic activities (UNWTO 2020) while contributing to the GDP and helping to create jobs, not to mention its capacity to promote the distribution of demand over time (to contrast seasonal peaks) and throughout a wider territory. In its "Recommendations on Tourism and Rural Development", the UNWTO states that the sustainability of tourism in rural areas will only prove successful if a comprehensive, inclusive planning strategy is adopted, and subsequently implemented, under a multi-action, multi-stakeholder participatory approach (UNWTO 2020).

In this sense, the adoption of holistic methodological frameworks that combine analytical, data-driven support for decision making with an all-encompassing narrative able to reinvent and communicate the territory makes it possible to manage and plan tourist offerings as an integrated component of the local ecosystem, in the interests of reconnecting rural hinterland areas with their wider territorial networks, so as to activate cultural and economic exchanges between local and external communities, as a way of valorising both rural culture and its natural treasures.

References

Antrop M (2004) Landscape change and the urbanization process in Europe. Landsc Urban Plan 67:1–4. https://doi.org/10.1016/S0169-2046(03)00026-4

ASIA (2022) Informative system on employment—employment. Accessed 20 February 2024. https://www.istat.it/en/archivio/282103

Barca F, Casavola P, Lucatelli S (2014) Strategia nazionale per le Aree interne: definizione, obiettivi, strumenti e governance. Documento tecnico collegato alla bozza di Accordo di Partenariato trasmessa alla CE il 9 dicembre 201. Accessed 20 February 2024. https://www.miur.gov.it/documents/20182/890263/strategia_nazionale_aree_interne.pdf/d10fc111-65c0-4acd-b253-63efae626b19

Basile G, Cavallo A (2020) Rural identity, authenticity, and sustainability in Italian inner areas. Sustain For 12(3):1272. https://doi.org/10.3390/su12031272

Dax T, Fischer M (2017) An alternative policy approach to rural development in regions facing population decline. Eur Plan Stud 26(2):297–315. https://doi.org/10.1080/09654313.2017.1361596

ENRD (2018) Smart villages revitalising rural services. EU Rural Rev:6–14

European Commission (2021) Staff working document. Accompanying the document Communication from the Commission to the European Parliament, the Council, the European economic and social committee and the committee of the regions. A long-term vision for the EU's rural areas—Towards stronger, connected, resilient and prosperous rural areas by 2040. https://eur-lex.europa.eu/legal-content/EN/ALL/?uri=SWD:2021:166:FIN. Accessed 21 Feb 2024

European Network for Rural Development (2020) Strategy for Inner areas. Italy. Working document. https://ec.europa.eu/enrd/sites/default/files/tg_smart-villages_case-study_it.pdf. Accessed 21 Feb 2024

Eurostat (2021) Urban-rural Europe—Introduction, https://ec.europa.eu/eurostat/statistics-explained/index.php?title=Urban-rural_Europe_-_introduction#Area_and_population. Accessed 21 Feb 2024

Garrod B, Wornell R, Youell R (2006) Re-conceptualising rural resources as countryside capital: the case of rural tourism. J Rural Stud 22(1):117–128

Haller A, Branca D (2022) Urbanization and the verticality of rural–Urban linkages in mountains. In: Sarmiento FO (ed) Montology palimpsest. Montology, vol 1. Springer, Cham. https://doi.org/10.1007/978-3-031-13298-8_8

ICOMOS (2017) ICOMOS-IFLA principles concerning rural landscapes as heritage. https://www.icomos.org/images/DOCUMENTS/Charters/GA2017_6-3-1_RuralLandscapesPrinciples_EN_adopted-15122017.pdf. Accessed 30 Nov 2021

International Labour Organization (2019) Economic diversification of rural economy. https://www.ilo.org/global/topics/economic-and-social-development/rural-development/WCMS_436223/lang%2D%2Den/index.htm. Accessed 21 Feb 2024

ISTAT (2018) Resident municipal population by age, sex and marital status. https://www.istat.it

ISTAT (2022) Resident municipal population by age, sex and marital status. https://www.istat.it

M'alikov'a L, Farrell M, McDonagh J (2016) Perception of marginality and peripherality in an Irish rural context. Quaest Geogr 35:93–105. https://doi.org/10.1515/quageo-2016-0037

NSIA (2013) National Strategy for Inner Areas. http://www.programmazioneeconomica.gov.it/2019/05/23/strategia-nazionale-delle-aree-interne/. Accessed 12 Feb 2024

Shane G (2005) Recombinant urbanism, conceptual modeling in architecture, urban design and city theory. Wiley-Academy Book, Chichester

Slee B (2020) Revitalising rural services through social innovation SIMRA. https://enrd.ec.europa.eu/sites/enrd/files/tg2_smart-villages_social-innovation_slee.pdf. Accessed 20 Feb 2024

Smith B, Ripp M (2020) Guidelines for sustainable cultural tourism a unique opportunity for change post Covid-19. In: Luger K, Ripp M (eds) World heritage, place making and sustainable tourism towards integrative approaches in heritage management. Studienverlag, Innsbruck

Sypok SI, Marinchenko TE (2023) Rural development as a factor of sustainable development. In: Beskopylny A, Shamtsyan M, Artiukh V (eds) XV international scientific conference "INTERAGROMASH 2022". INTERAGROMASH 2022, Lecture notes in networks and systems, vol 575. Springer, Cham. https://doi.org/10.1007/978-3-031-21219-2_253

UN Statistical Commission (2020) A recommendation on the method to delineate cities, urban and rural areas for international statistical comparisons. United Nations, New York. https://unstats.un.org/unsd/statcom/51st-session/documents/BG-Item3j-Recommendation-E.pdf

United Nations (2021) World social report 2021: reconsidering rural development. UNDESA, New York. http://www.un.org/development/desa/dspd/wp-content/uploads/sites/22/2021/05/World-Social-Report-2021_web_FINAL.pdf

UNWTO (2020) Recommendations on tourism and rural development—a guide to making tourism an effective tool for rural development. UNWTO, Madrid. https://doi.org/10.18111/9789284422173

van Vliet J, Verburg PH, Grădinaru S, Hersperger AM (2019) Beyond the urban-rural dichotomy: towards a more nuanced analysis of changes in built-up land. Comput Environ Urban Syst 74:41–49. https://doi.org/10.1016/j.compenvurbsys.2018.12.002

Wachter S (2001) Les mutations de l'accessibilité: risques et chance pour les politiques d'aménagement. Notes CPVS 15:5–22. https://temis.documentation.developpement-durable.gouv.fr/docs/Temis/0073/Temis-0073018/19966.pdf

Chapter 4
The Decarbonization of Urban Districts by 2050 in Mediterranean Area: Green and Grey Solutions

Fabrizio Tucci and Marco Giampaoletti

Abstract The climate changes we are currently witnessing, accompanied by massive emissions of greenhouse gases on a global scale, are largely anthropogenic in nature, and increasingly reinforce awareness of the need for rapid and timely intervention to contain new carbon emissions into the atmosphere while mitigating their effects.

This chapter outlines possible strategies, methodologies, and actions to contribute to the decarbonization of urban districts. Beginning with a detailed review of the literature forming the basis of a structured methodology, a dual field of application has been defined: (1) "green" strategies and actions, through the identification and selection of tree species present in the Mediterranean area with high carbon absorption capacities, and (2) "grey" strategies, actions, and solutions, through the adoption, at various scales, of available technologies and processes for carbon storage (in building envelopes, open space pavements, and underground). These fields of inquiry have been tested in an applied case study (the ERP district of San Basilio, Rome), proposing different intervention scenarios. The contribution aims to define a new original methodological-applicative approach to support the design and meta-design phases in the regeneration of urban districts, aimed at offering concrete responses to contemporary challenges in the field of climate change adaptation and mitigation. Moreover, the research targets designers, architects, and planners and aims to establish a knowledge base and decision-making support through a framework of best practices and actionable measures in the field of construction.

Keywords Decarbonization · Green and grey solution · Urban reforestation · Natural carbon sinks

F. Tucci (✉) · M. Giampaoletti
Department of Planning, Design, Technology of Architecture, Sapienza University of Rome, Rome, Italy
e-mail: fabrizio.tucci@Uniroma1.it; marco.giampaoletti@uniroma1.it

A. Battisti et al. (eds.), *Urban-Rural Dialogue in Green City Design*, Designing Environments, https://doi.org/10.1007/978-3-031-88318-7_4

4.1 Introduction

In the face of climate change and the growing European energy crisis, recently exacerbated by rising supply costs, urgent policies and actions are needed to make city dwellers more resilient and responsible. Such strategies can no longer be postponed in light of the current impacts resulting from sudden economic, social, and environmental changes. Compared to the period of 1850–1900, the average global temperature has increased by 1.1 °C (IPCC 2023). This increase was mainly caused by human activities as well as the effects that these have produced on the environment. Global average temperature increased both on land (+1.59 °C) and in the oceans (+0.88 °C) (IPCC 2023).

The current concentration of carbon dioxide (412 ppm) is the highest in 2 million years, while concentrations of methane (1866 ppb) and nitrous oxide (332 ppb) are the highest in 800,000 years (OMM 2024). Considering the period between 1850 and 2019, it is possible to observe that 42% of the total greenhouse gases of anthropogenic origin have been emitted in the last 30 years (IPCC 2023). As concern grows, the urgency of deep decarbonization increases, especially in cities. In a scenario which underlies a constant increase in demand for the redevelopment of existing building stock, the need to define decision-making and operational support tools capable of prefiguring and verifying the effectiveness of the transformations clearly emerges.

The need for actions aimed at improving the energy performance of the existing building stock is evident. In the European Union more than 40% of energy consumption is attributable to the operation of buildings in the residential and tertiary sectors. The survey on the existing building stock, of which over 65% predates the first law on energy saving (ISTAT 2021), analyzes European commitments aimed at reducing climate-altering gas emissions, deriving mainly from energy consumption for heating and cooling. The urban and architectural regeneration project therefore takes on a central role in central urban management, proposing new design tools capable of increasing the quality of the final product in terms of energy consumption and resources.

In particular, the contribution focuses on the application, on an experimental basis, in a public residential building district of Rome, of solutions capable of capturing, removing, and storing carbon emissions. These solutions may be described by two macro-concepts: "green" and "grey." While some of these solutions are already in use, others have the potential to be expanded technologically; others, still, appear only theoretical at the present time. In the research, different solutions are analyzed by estimating their performance impact on net carbon storage. Applicability and interaction are then verified, evaluating the limits and opportunities from an economic point of view, as well as scientific dissemination for the main players in the sector. In response to these problems, there is an urgent need for massive decarbonization, focusing on the areas of the world most difficult to decarbonize, with the aim of removing large quantities of carbon dioxide through monitored, reported and verifiable processes (Climeworks 2023). As numerous studies emphasize, the "green" and "grey" aspects are often treated separately, without understanding their potential interactions in terms of economic, social, and environmental benefits, thus effectively limiting a unitary vision (Kuittinen 2021a, b; Ariluomaa et al. 2021; Amiri et al. 2020).

The AEC sector therefore becomes a central pivot of this change, strongly claiming a new integrated and transversal approach to the entire planning, design, and production process, fully integrated with the innovative scope of the digital transition (Abdolhosseini Qomi et al. 2016).

Alongside the more general theme of urban adaptation policies, attention is increasingly focusing on Nature-Based Solutions (NBS), from a 'green' perspective, effective in creating more resilient urban contexts, while guaranteeing a multiplicity of social benefits, environmental and economic (Catalano et al. 2021). Nature-based urban projects, (European Commission 'Nature-Based Solutions and Renaturing Cities' 2015), implies measures inspired, supported, and replicated by natural mechanisms that allow for increasing resilience of urban contexts while ensuring greater attractiveness and opportunities for urban development, as well as quality of life issues linked to urban growth (Tucci 2021).

The incentive strategy for green infrastructure actions and solutions is part of a broad scientific debate concerning the development of integrated methodologies and tools for dissemination, with the aim of promoting the sustainability of urban transformations in a perspective of profound renewal (Andreucci 2017).

Among the urban regeneration measures that aim to decarbonize neighborhoods, with multiple solutions based on a main strategic axis defined as "green," afforestation and reforestation practices may be included, in addition to environmental technology solutions (NBS) capable of offering responses to the needs of the problem of loss of Natural Capital and quality of life increasingly linked to urban growth. This is the meaning of the European Commission's FitFor55 climate plan, which aims to improve the quality, quantity, and resilience of green areas with the reforestation of European forests through 3 billion trees to be planted by 2030, and specific national objectives for the expansion of natural carbon (Bionova 2018).

The design use of these solutions is mostly intended for development mechanisms whose effects are known, but increasingly new situations are explored by imitating the way in which non-human organisms and communities deal with extreme events (Antonini and Tucci 2017). Therefore, the adoption of actions aimed at targeted and planned urban reforestation in Mediterranean climatic conditions determines climate mitigation effects with a calculable and verifiable CO_2 subtraction, a widespread improvement of the local microclimate, a clear increase in ecosystem services and a general increase of natural resources; these results can be achieved if the valorization of the tree and shrub masses occurs using autochthonous species, native to the area and with a high environmental mitigation capacity.

The research presented in this chapter considers, in parallel, another source of greenhouse gas emissions associated with buildings, which is embodied CO_2, i.e., the set of greenhouse gas emissions associated with construction, and particularly those deriving from the extraction, transport and production of materials, which represent the main share of the total carbon embodied in buildings. With the aim of monitoring and mitigating this significant share of emissions, which represent a total of 11% of global carbon emissions (Pomponi et al. 2018; WGBC 2019), the research identifies, for the 'grey' strategies, intervention scenarios that on the one hand, they use low embodied carbon materials, for example using wood or plant fibers for natural thermal insulation to exploit the carbon emissions stored in species suitable for the production of such building materials.

The solutions defined as 'grey' finally take into consideration the systematization of the technologies and processes available today for the storage of CO_2, for example the CCS (Carbon Capture and Storage), the CCU (Carbon Capture and Utilization), the DAC (Direct Air Capture) (Global CCS Institute 2022). As regards these technologies, the useful elements for their concrete integration with natural solutions and their potential applicability in the built context are examined, through the identification of technologies and products that can be used today safely and effectively, so that they can contribute to achieve the planned objectives.

4.2 Methodology and Operational Phases of the Research

The analysis develops in three operational phases. The first phase focuses on the quantification of CO_2 emissions for the areas of investigation based on the actual state of the area being tested. The second phase analyzes the advanced design options in view of the "green" and "grey" directions. The third phase involves an evaluation of the issues examined according to two intervention scenarios (the first, an ordinary redevelopment, the second, an advanced redevelopment through the use of innovative materials). An important support in this regard is the Tree Census carried out by the Municipality of Rome in June 2016 for each individual municipality. Starting from this census, the tree and shrub masses present in the area subject to the experimental intervention were censused, proposing a qualitative-quantitative index that considered not only the botanical typology, but the historical, cultural and landscape parameters as well. This knowledge base, combined with direct surveys, represents a more complete panorama of the tree species present in the Mediterranean basin and with the highest carbon absorption capacity. This has allowed the creation of a database of the main tree and shrub species present in the Metropolitan City of Rome, with a high capacity for environmental mitigation and absorption of climate-altering gases; this database can be made available to local authorities, and may be of use to private citizens and businesses.

This methodological tool aims to establish the CO_2 absorption and storage values for each analyzed species. During the preparation of the database, a "qualitative-quantitative taxonomic conservation index" (Taxonomic Storage Index—TS) was introduced to raise awareness about the use of native tree species with high environmental mitigation capacity, which are more sustainable in terms of environmental impact reduction (improvement of the thermo-hygrometric system of the plant area and selective filtering capacity of urban air pollutants).

The index is based on the mathematical relationship between a species' ability to absorb carbon from the atmosphere (C) and its exotic status (E):

$$TS = \frac{C}{E}$$

The following formula, with the index tending to 1, involves the adoption of tree species with greater environmental mitigation capabilities, carbon absorption, and a higher degree of exoticness (E). When the values tend toward 0, this choice becomes unsustainable and inapplicable for the local urban context. The index has been structured and organized based on a detailed thematic literature review covering the period from 2007 to 2021 (Pignatti et al. 2017; Metropolitan City of Rome 2021), which is essential for understanding the actual environmental mitigation capacity available in the metropolitan area under analysis and is preparatory to quantifying the carbon absorption capacity that such a heritage is capable of. During the winter and summer climatic seasons, campaigns were conducted to survey the existing tree and shrub vegetation, thereby enriching the available and consulted literature. This phase led to the creation of a detailed database using Office 2020 suite, in which the scientific name of each species was cataloged (also known as the "binomial nomenclature," a formal system for naming species of living organisms, including trees). Each tree was given a name consisting of two parts: genus and species, along with photographic documentation of the leaf structure, flower, and fruit, the number of individuals surveyed within the analyzed area, and the exotic status (Galasso et al. 2018), indicating whether the species is native (originated and evolved in the area where it is found) or alien (introduced due to human intervention outside its natural range, including those that do not form stable populations). Specifically, for alien species, a further cataloging was performed for archeophytes (introduced before 1942) and neophytes (introduced after 1942). For alien species, the casualness of spontaneous development and reproduction was classified, as well as their degree of naturalization (the ability to form stable, independent populations without active human involvement) and invasiveness (the potential to outcompete native and indigenous tree species in each habitat without human intervention, rapidly reproducing on a large scale). This enabled classification according to biological form (or Raunkiær scale, a category of plants that possess a specific morphological adaptation to environmental factors, based on the position of the bud during unfavorable seasons), divided mainly into five groups: phanerophytes, chamaephytes, hemicryptophytes, geophytes, and therophytes. The first four groups include perennial species, while the last group consists of species with a shorter life cycle. The database also includes a section on the optimal potential use of the analyzed species (craft, food, forestry, melliferous, ornamental, etc.), as well as its capacity for carbon storage and absorption from the atmosphere (in tons per year, over 20 years, for a species of an average age of 25 years) (Tuscany Region 2018). Subsequently, the analyzed species were reclassified according to a simplified corotype classification based on the classification provided by ActaPlantarum and to homogenize the geographic macroareas of origin (Table 4.1).

Table 4.1 On the right of the table the Chorotype according to the classification reported by Actaplantarum, on the left the reclassified Chorotype (Tucci, Giampaoletti 2023)

Chorotype of Actaplantarum	Reclassified chorotype
North America	Alien
South America	Alien
Asia	Alien/Eurasian
Western Asia	Eurasian
East Asia	Alien
Australia	Alien
Europe	Mediterranean
Southern Europe	Mediterranean
Asian European	Eurasian
European Asian, Eurasian, North African	Eurasian
European Caucasian	European-Caucasian
Eurosiberian	Eurosiberian
Macaronesia	Mediterranean
Mediterranean	Mediterranean
Naturalized	Eurasian/Alien/ Euro-Caucasian
Countries in the tropical belt of Africa and Asia	Alien

As for the "gray" solutions, the phases of the methodology adopted in the project for the redevelopment of existing buildings and their associated outdoor spaces followed a detailed process of constructive analysis of the existing building to characterize it in terms of performance, technology, and materials. This included targeted identification and quantification of the main materials present, as well as a possible estimate of the carbon embedded in the existing materials based on relevant databases, an estimate (in volume and weight) of the materials removed from the existing building and their related embedded carbon according to an estimate of the material needs (volume) for the intervention. Subsequently, an identification of recoverable materials from the demolition phase was carried out, calculating the portions of recyclable/reusable materials both on-site and off-site; this was preparatory to the selection of low-carbon-emission materials and innovative materials with carbon storage or direct capture capabilities from the atmosphere.

Finally, the carbon balance of the redeveloped buildings was defined, identifying the portion of emissions avoided thanks to the reuse of the existing buildings, the recycling of materials, and the use of low-carbon-emission materials or those with CO_2 storage capabilities. These results served as the basis for calculating the share of emissions that can be captured over time through the use of appropriate materials.

4.3 Experimentation. The Case of the PdZ of San Basilio in Rome

The public housing district of San Basilio, designed by M. Fiorentino between 1951 and 1955, is located in Municipality IV of the Municipality of Rome, within which there are about 4680 ERP housing units (Fig. 4.1) with a total area of over 130,000 square meters, half of which are classified as open and permeable spaces.

These accommodations are managed 70% by ATER—Territorial Agency for Public Residential Building (formerly IACP—Istituto Autonomo Case Popolari) and 30% by the Municipality, located precisely in the former PDZ 02V—San Basilio.

The building typologies present are characterized by linear slat buildings, with almost homogeneous heights of up to 7 floors and the conformation of the building based mainly on prefabricated 'closed cell' and overlapping structures, with floors made of predalles slabs that define the internal partitions of the apartments (Fig. 4.2). The embodied carbon quantified using specific databases for the differentiation of the individual building materials surveyed is equal to 54,849,066 $KgCO_{2eq}$.

The 86,603 m^2 of open spaces of the district are characterized by large green areas that are mostly degraded (53,473 m^2) and by roadways and residents' car parks. Bituminous asphalt was used for the streets and sidewalks, blocks of leucite and basalt in the areas pertaining to the buildings and in the four central courtyards

Fig. 4.1 Aerial photo of the PdZ of San Basilio (PDTA Paper Library, 'Sapienza' University of Rome, 2014)

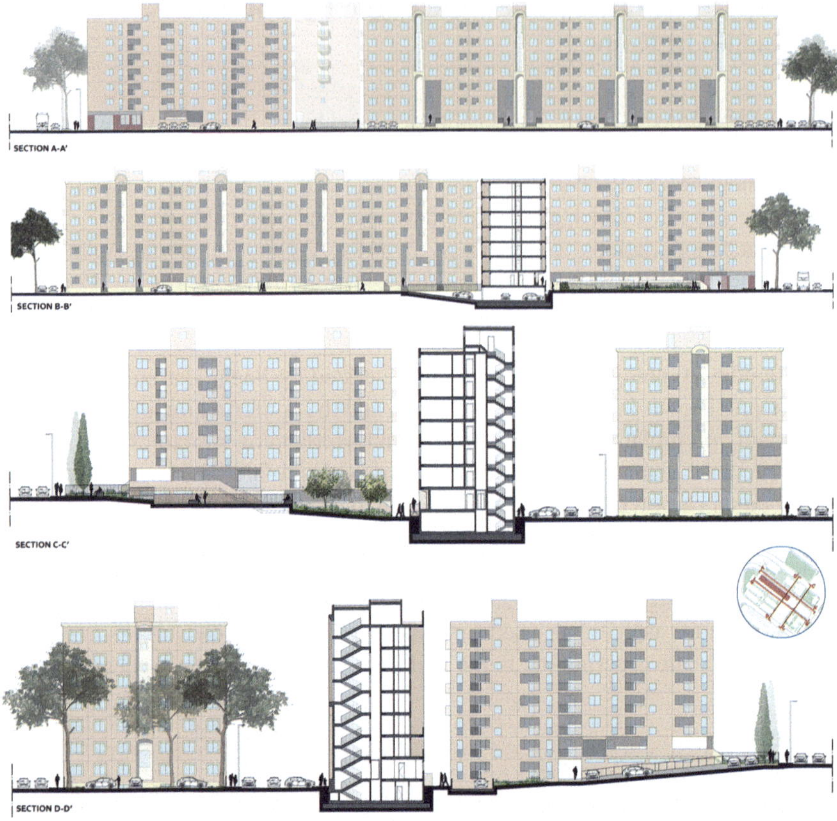

Fig. 4.2 Perspectives of a typical building in the San Basilio PdZ (Tucci, Giampaoletti 2023)

of the district, redeveloped in 2008 with the inclusion of play spaces in anti-trauma flooring, small equipment for stop and a small kiosk bar. The pedestrian paths are made of concrete tiles with a grit finish (Fig. 4.3). The overall embodied carbon amounts to 235,392,400.74 $KgCO_{2eq}$.

The study and analysis of the urban regeneration aspects of the PdZ included the identification and analysis of the tree and shrub species present in the area. This led to the cataloging of over 20 species of trees and shrubs, highlighting a predominance of *Ailanthus altissima*, *Cedrus atlantica*, *Robinia pseudoacacia*, as well as herbaceous and tree species that are alien and invasive to the area under analysis, totaling 374 individuals with medium to tall trunks, with a total carbon sequestration of 360,576 Kg CO_2eq/year (Table 4.2).

The species surveyed are predominantly non-native (over 64%) with low environmental sustainability, according to their exoticism status, predominantly indigenous with about 36%. Following with the same presence percentage are invasive neophytes, naturalized neophytes, and random neophytes, each with 14%. The

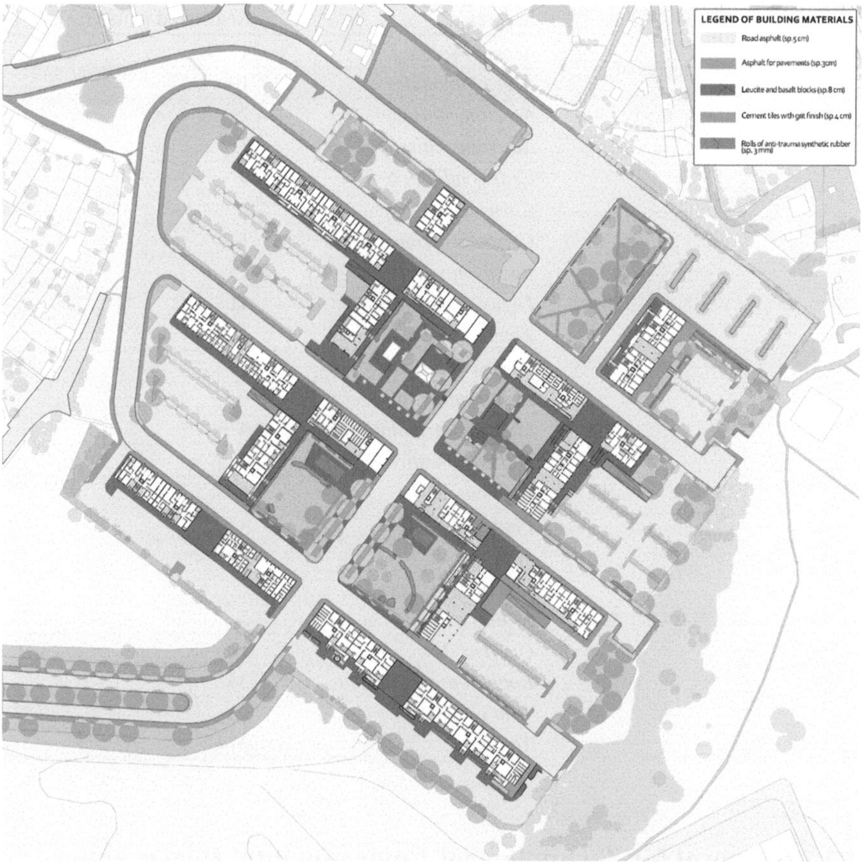

Fig. 4.3 Census of building materials present in the current state (Tucci, Giampaoletti 2023)

presence in this area is concluded by naturalized neophytes and random archaeophytes, with a coverage percentage of 9%.

The total capacity to absorb CO_2 from the atmosphere of the current tree masses and green spaces has been estimated at over 360,000 $KgCO_2eq$/year, classified as having a Medium-High and High level of capacity. The species with the highest potential for carbon absorption and storage is 'Pinus pinea', a naturalized and non-spontaneous archaeophyte species in the Metropolitan City of Rome (Fig. 4.4).

The index applied to the recorded species shows a homogeneous distribution of frequency with a maximum value close to 1, a minimum value of 0.002, and an average value of 0.163. The Average Taxonomic Storage Index (TSm) of the analyzed and registered species is 0.147, 43% higher than the average recorded for the census of tree and shrub species present within the municipal territory.

Table 4.2 Classification and numerical quantification of the tree and shrub species of the PdZ of San Basilio (Tucci, Giampaoletti 2023)

Species	Registered number
Acacia dealbata	4
Acer platanoides	10
Acer saccharinum	21
Ailanthus altissima	31
Cedrus atlantica	9
Citrus lemon	2
Cupressus sempervirens	17
Eucalyptus camaldulensis	12
Fagus sylvatica	41
Junglans regis	11
Laurus nobilis	70
Liriodendron tulipifera	6
Magnolia grandiflora	4
Malus domestica	3
Olea europea	6
Pinus nigra	4
Pinus pinea	30
Platanus hyspanica	6
Prunus avium	5
Quercus ilex	10
Robinia pseudoacacia	34
Tilia cordata	38
Total	**374**

4.4 Proposal for Technological-Environmental Regeneration Aimed at Removing Carbon and Increasing Environmental Quality

The redevelopment through 'grey' strategies of the building stock and open spaces has examined two options. The first, with standard materials that are usually used in renovations according to current state legislation, is therefore based on the maintenance of the load-bearing structure in reinforced concrete and on an application of building materials such as blocks and insulators for horizontal closures for the reconstruction of the external envelopes, of roofing and the replacement of fixtures, capable of increasing and improving thermal and energy performance.

In relation to external spaces, the working hypothesis envisages an increase in pedestrian areas by introducing basalt and leucite slabs, and new play areas (Fig. 4.5).

For the materials subject to demolition such as concrete, bricks and plasters, there is in situ recovery with a mobile plant for the treatment of inert waste and the reuse of the recycled aggregates for the reconstruction of the driveway road

Fig. 4.4 Summary of Native and Exotic Flora (Tucci, Giampaoletti 2023)

foundations and the environmental and landscape arrangements of the urban district. As regards materials coming from the replacement of fixtures such as aluminum and glass, off-site management is considered for further use after industrial treatment.

The mix of strategies proposed in the first redevelopment option provides for a quantification of embodied carbon equal to 142,570,013.88 $KgCO_{2eq}$.

The second option involves the use of experimental materials, often not yet on the market, carbon free or with extremely low emissions in all phases of production and processing and in some cases capable of actively contributing to the absorption

Fig. 4.5 Innovative urban regeneration proposal (Tucci, Giampaoletti 2023)

of carbon from the atmosphere (Fig. 4.6). Specifically, interventions focus on maintaining the existing load-bearing structure, with the redefinition of vertical closures, through the introduction of large glass surfaces with low-emissivity windows, passive bioclimatic devices such as atriums to the south, buffer spaces to the north and projecting bioclimatic greenhouses which determine an increase in volumes in line with current regulations, improving the adaptability and flexibility of the accommodation in order to have greater environmental comfort, reducing energy consumption for heating and cooling. The external infills are made up of external insulation walls in Carbfix blocks, derived from the process of capturing CO_2 in the atmosphere using DAC (Direct Air Capture) technology, liquid injection into the

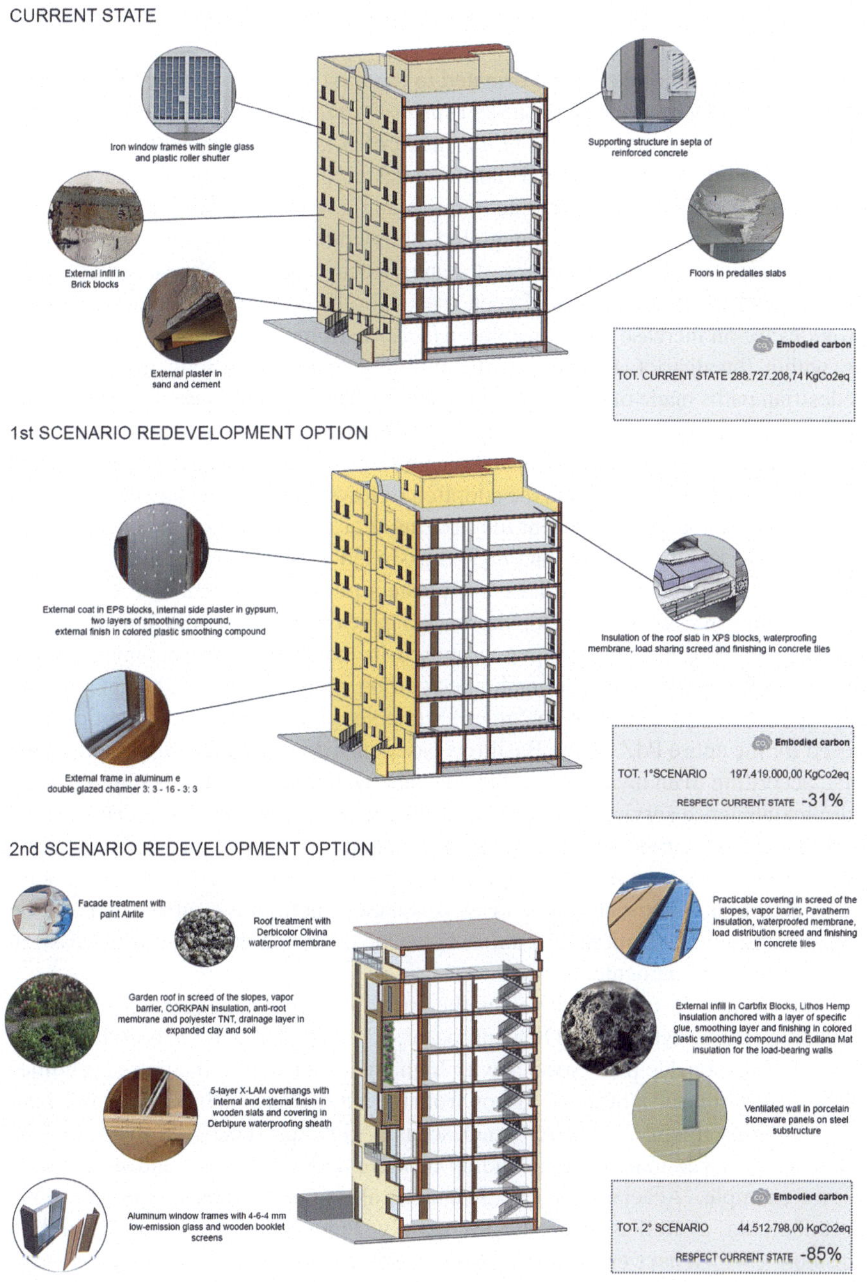

Fig. 4.6 Graphic representation of the materials present in the current state and those introduced for the first and second redevelopment scenario's proposal (Tucci, Giampaoletti 2023)

subsoil consisting mainly of basaltic rocks and subsequent mineralization, thus generating a building material with carbon storage capacity.

The vertical external closures oriented to the west include ventilated walls made of porcelain stoneware panels on a steel substructure, while the overhangs are made up of bioclimatic greenhouses and are made of 5-layer X-Lam with internal and external finishing in treated wood slats. The design of the horizontal closures, insulated with Corkpan sheets, provides practicable coverings, garden roofs and surfaces for the introduction of integrated photovoltaic panels with a Derbicolor Olivina finish, the latter experimental material with high CO_2 absorption capacity. The project for the redefinition of open spaces for the present redevelopment option provides for an increase in permeable surfaces, a strong reduction in vehicular traffic within the district, favoring gentle mobility, with the introduction of cycle-pedestrian paths made of cement conglomerate I. Hydro Drain with high absorption and storage capacity of polluting particles within its molecular structure.

The design of convivial spaces and recreational areas to host outdoor events and commercial activities considers the adoption of Revstone flooring, leucite and basalt blocks and photocatalytic cement tile flooring. For the materials subject to demolition, a recovery rate of more than 88% is assumed with in situ treatment of aggregates coming from concrete, bricks, and plasters for the construction of screeds and external flooring, as well as, for the excess portion, for the arrangement district landscaping. The mix of strategies proposed in the second redevelopment option provides for a quantification of embodied carbon equal to 15,021,500 $KgCO_{2eq}$. As regards redevelopment through 'green' strategies, the greening intervention proposed for the entire PdZ of San Basilio is based on a double regeneration option; the first, according to an invariance of the vegetal cover present, strengthening the periodic maintenance service and attention to the most delicate essences. The second option instead focuses on a qualitative and quantitative project increase in existing plant masses, native and with a high mitigating effect on the environment, potentially eliminating non-native and invasive species, and, on the other hand, on the recovery, functionalization, and valorization of open spaces, with an increase in their overall environmental quality.

The scenario in question involves the strengthening of resilient native species including 'Fagus sylvatica', 'Olea europaea', 'Pinus pinea' and the introduction of five other sustainable plant species with high environmental mitigation capabilities in the absorption and storage of carbon from the atmosphere such as 'Pinus halepensis', 'Populus tremula', 'Quercus pubescens', 'Quercus suber'. Furthermore, the eradication of 128 individuals including invasive and weedy trees and shrubs such as, for example, 'Acacia dealbata', 'Ailanthus altissima', 'Junglas regia', 'Robinia pseudoacacia' as well as shrub species, mainly 'Pinus pinea', too close to building structures, which have created, over time, problems of natural ventilation and shading in the inspections carried out in the summer and winter periods. Overall, 587 new tree and shrub individuals are introduced in the degraded, recovered, and redeveloped areas (32 genera for a total quota of 961 individuals) for a total carbon absorption of 506,816 $KgCO_{2eq}$/year. The 32 species present in this redevelopment scenario are almost exclusively predominantly native (94%, 30 species) with a significant increase in environmental sustainability demonstrated, according to the State of Exoticity, by the establishment of predominantly indigenous species (91%, 29 species), followed by random neophytes (3%, 1 species), naturalized neophytes (3%, 1 species) and naturalized archaeophytes (3%, 1 species).

4.5 Discussion of the Results

The adoption of the strategies described above has defined project lines that can be verified by quantifying the effects they introduce in the area being tested, in terms of unitary embodied carbon performance, absorption and storage of carbon from the atmosphere. The strategy of increasing the tree and shrub heritage, in the second 'green' redevelopment option, with the introduction of new tree species, native and with a high environmental mitigation capacity, has resulted in an increase of +183% in carbon absorption, while 8% is absorbed using some specific innovative and experimental 'grey' materials (Fig. 4.7).

In terms of embodied carbon, compared to construction in the actual situation, the scenarios examined in the 'grey' strategies determine a reduction of 32% in the first option and 77% in the second option (Fig. 4.8).

The absorption capacity of CO_2 from the atmosphere exercised by the arboreal heritage of the area, following the operations planned and simulated according to the botanical and biogenetic parameters reported in the Methodology, appears to be of a high level overall, thus improving the absorption performances compared to those state of things. It should be noted that the species with the highest potential for carbon absorption and storage is 'Pinus halepensis', an autochthonous and spontaneous entity of the Metropolitan City of Rome and with a high capacity for environmental mitigation, which not surprisingly was among the introduced species in the urban reforestation project of the area. The TS index applied for the species present in this scenario presents a homogeneous distribution frequency with the maximum

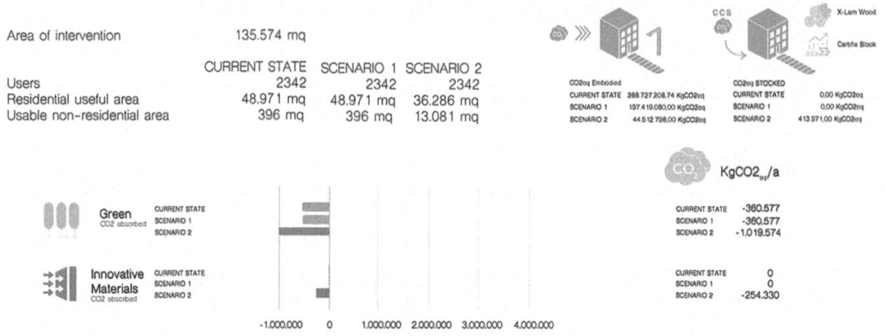

Fig. 4.7 Final balance of CO_2 emissions according to the green and gray strategies (Tucci, Giampaoletti 2023)

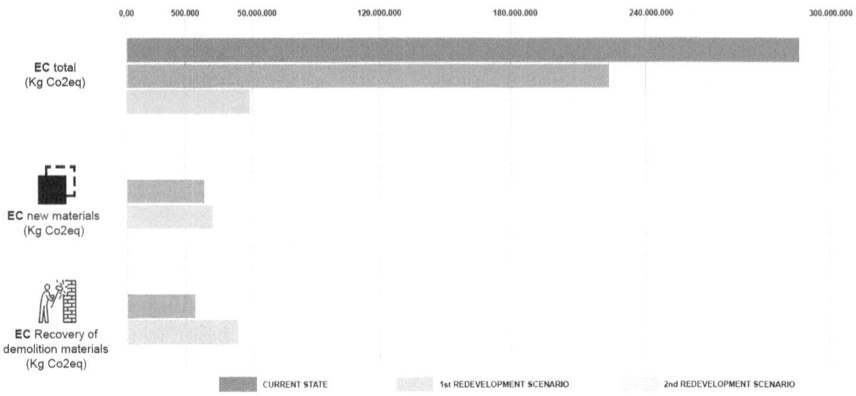

Fig. 4.8 Final balance of the total embodied carbon of the new materials introduced and of the recovery of demolition materials, according to the State of Fact, first and second Redevelopment Scenario (Tucci, Giampaoletti 2023)

value equal to 1, a minimum value equal to 0.008 and a TSm index equal to 0.188, increasing the environmental capacity of carbon absorption in relation to the exotic status, compared to the actual state.

This study has highlighted important solutions and design strategies in the field of 'green' and 'grey' solutions aimed at increasing environmental well-being in a peri-urban neighborhood such as the densely urbanized city of Roma Capitale, proposing targeted intervention scenarios to increase the capacity of storage of carbon from the atmosphere through urban reforestation actions with local and native species with high resilience capabilities compared to the alien species present on site.

The increase in the number of tree and shrub species registered and implemented in the database presented will also contribute to refining and improving the TS index as it is applicable in every national and, in perspective, European (including global) context and comparable at an international level by comparing multiple areas, even distant from each other, using the TSm index. This indicator may also find application potential with respect to the forest regeneration capacity present in a specific area aimed as a planning intervention for a specific neighborhood which needs such beneficial actions.

4.6 Conclusion

The methodological structure described in this chapter aims to experimentally analyze the application fields related to green and grey solutions capable of decarbonizing urban districts through carbon capture, removal, and storage. This design approach has defined the strategies to follow in order to assess the effects of the design choices adopted in the two intervention scenarios. Increasing the qualitative gradient of permeable and green areas thus becomes of fundamental importance in terms of climate change adaptation and mitigation, contributing, with appropriate measures, to a global decarbonization process. The time to adopt these solutions is slowly running out, and the commitments made by various countries worldwide are currently far from the set goals, with some threatening to withdraw from global climate agreements. Furthermore, the difficulty and complexity of implementing methodologies, strategies, and guidelines aimed at reducing CO_2 emissions—both green and grey—globally are highlighted by the lack of a non-harmonized legislative framework and a national, European, and global funding plan that is inadequate for the challenges we are called to face. This research has highlighted limitations in specific aspects related to green and grey solutions aimed at contributing to the decarbonization of urban and peri-urban districts, often due to their application in states with different geopolitical situations or to alternative solutions that are often more established and economically advantageous. Actions related to green strategies increasingly benefit from technological solutions that are more accessible and cost-effective; however, the lack of knowledge among sector actors and the effective dissemination of such knowledge, particularly at the national level, often leads to the use of low-quality tree and shrub species, with ages incompatible with the highly urbanized context of a city, with low environmental mitigation capacities, and costs that are often higher than similar, resilient, native species. Actions and solutions affirmed in grey strategies highlight the research limits regarding the use of these technologies in relation to quantified and defined climate changes. Technologies applicable on urban and industrial scales, such as Carbon Capture and Storage (CCS) and Direct Air Capture (DAC), are among the most debated points in the global climate policy discussions, considered ineffective if distant from industrial applications, or unreliable because they are not yet used, particularly on a national scale, due to the extensive areas with high seismic risk.

References

Abdolhosseini Qomi MJ, Noshadravan A, Sobstyl JM, Pellenq RJ-M, Ulm F-J, Gonzalez MC, Toole J, Ferreira J (2016) Data analysis to simplify thermal efficiency planning in cities. J R Soc Interface 13:2. https://doi.org/10.1098/rsif.2015.0971

Amiri A, Ottelin J, Sorvari J, Junnila S (2020) Cities as carbon sinks—classification of wooden buildings. Environ Res Lett 15:09407615. https://doi.org/10.1088/1748-9326/aba134

Andreucci MB (2017) Designing green infrastructure—technologies, values and tools for urban resilience. Wolters Kluwer, Milanofiori Assago

Antonini E, Tucci F (eds) (2017) Architecture, city and territory towards a green economy. Building a green economy manifesto for the architecture and city of the future. Ambiente Editions, Milan

Ariluomaa M, Ottelinb J, Hautamäkia R, Tuhkanenc E, Mänttärid M (2021) Carbon sequestration and storage potential of urban greenery in residential yards: a case study from Helsinki. Urban For Urban Green 57:126939. https://doi.org/10.1016/j.ufug.2020.126939

Bionova, One Click LCA (2018) The Embodied Carbon Review Carbon reduction embodied in over 100 regulations and classification systems globally. Bionova Ltd. Available via DIALOG. https://www.oneclicklca.com/embodied-carbon-review/. Accessed10 November 2023

Catalano C, Andreucci MB, Guarino R (2021) Urban services to ecosystems: green infrastructure benefits from the landscape to the urban scale. Springer, Berlin

Climeworks (2023) Air capture summit 2023. Available via DIALOG. https://climeworks.com/uploads/documents/climeworks-direct-air-capture-summit-2023%2D%2D-ceo-opening-keynote-for-publication.pdf. Accessed 27 October 2023

Galasso G et al (2018) An updated checklist of exotic vascular flora in Italy. Plant Biosyst 152:556–592

Global CCS Institute (2022) Global status of CCS 2022, Melbourne VIC 3000, Australia. Available via DIALOG. https://status22.globalccsinstitute.com/. Accessed 4 November 2023

IPCC—Intergovernmental Panel on Climate Change (2023) AR6—climate change 2023—summary report, Switzerland. Available via DIALOG. https://report.ipcc.ch/ar6syr/index.html. Accessed 7 November 2023

ISTAT (2021) Italian statistical yearbook 2021. Streetlib, Rome Available via DIALOG. https://www.esteri.it/wp-content/uploads/2023/12/STATISTICAL_YEARBOOK_2021.pdf. Accessed 13 May 2024

Kuittinen M, Hautamäki R, Tuhkanen E, Riikonen AM (2021a) Environmental product declarations for plants and soils: how to quantify carbon uptake in landscape design and construction? Int J Life Cycle Assess 26:1100–1116. https://doi.org/10.1007/s11367-021-01926-w

Kuittinen M, Zernicke C, Slabik S, Hafner A (2021b) How can carbon be stored in the built environment? A review of potential options. Archit Sci Rev 66:91–107. https://doi.org/10.1080/00038628.2021.1896471

Metropolitan City of Rome Capital (2021) Arboreal budget. Available via DIALOG. https://www.comune.roma.it/web/it/scheda-servizi.page?contentId=INF70550. Accessed 9 November 2023

OMM (2024) State of the global climate 2023, WMO, Geneva. Available via DIALOG. https://library.wmo.int/records/item/68835-state-of-the-global-climate-2023. Accessed 13 May 2024

Pignatti S, La Rosa M, Guarino R (2017) Flora of Italy, vol 1–4. Bologna, Edagricole

Pomponi F, De Wolf C, Moncaster A (eds) (2018) Embodied carbon in buildings. Measurement, management, and mitigation. Springer, Berlin

Tucci, F (2021) Climate responsive envelopes—Frontiers of technological and performance innovation in building envelope design. Frontiers of technological and performance innovation in building skin design. Altralinea Edizioni, Florence

Tuscany Region (2018) Regional plan for ambient air quality. Guidelines for planting specific tree species for the absorption of nitrogen dioxide, fine particulate matter and ozone. Available via DIALOG. https://www.regione.toscana.it/documents/10180/23809530/Allegato-1-LG-Piantumazione-agg.+31+dic-2021.pdf/da42f961-13e2-d2a6-05e3-ec35bfe32dad?t=1644563443162. Accessed 15 November 2023

World Green Building Council (2019) Bringing embodied carbon upfront. Coordinated action for the building and construction sector to tackle embodied carbon, WGBC. Available via DIALOG. https://www.worldgbc.org/sites/default/files/WorldGBC_Bringing_Embodied_Carbon_Upfront.pdf. Accessed 5 November 2023

Chapter 5
Ecological and Landscape Networks as Strategic and Structural Components of Inter-municipal Planning

Carmela Mariano

Abstract This chapter is based on the author's research activities on models of government of metropolitan territories and the relations between urban planning and emerging forms of supra-municipal aggregation, with reference to metropolitan areas (Italian and European) in which bottom-up initiatives of inter-municipal cooperation (bottom-up model) and supra-municipal government bodies (top-down model) coexist. The chapter proposes a reflection on the theme of the inter-municipal dimension, as a minimum territorial reference for the implementation of environmental and landscape protection and enhancement functions, investigating the role of connective networks of green and natural and anthropic components within inter-municipal planning processes aimed at a strategy of metropolitan rebalancing and integrated urban regeneration. The chapter presents one of the European case studies investigated in the research and aims to identify possible theoretical-methodological and operational references for an innovative planning system that adopts an inclusive conception of the territory for unitary knowledge and design of the different components. Within this framework, the chapter focuses on the experience of the Scheme of Territorial Coherence (SCOT) of the Grenoble Urban Region, in which the system of networks of natural, semi-natural and anthropic components represents the reference framework for the implementation of urban regeneration strategies and metropolitan territorial rebalancing and, at the same time, the invariant for the development and reconfiguration of the physical and socio-economic assets of the vast area territory, using the concepts of nature and landscape as catalysts for regeneration and urban well-being.

Keywords Ecological networks · Inter-municipal planning · Urban and environmental regeneration · Sustainability · Landscape

C. Mariano (✉)
Department of Planning, Design and Technology of Architecture (PDTA), Faculty of Architecture, Sapienza University of Rome, Rome, Italy
e-mail: carmela.mariano@uniroma1.it

A. Battisti et al. (eds.), *Urban-Rural Dialogue in Green City Design*, Designing Environments, https://doi.org/10.1007/978-3-031-88318-7_5

5.1 Strategies and Goals for Ecological Transition and Sustainability

The Sixth Assessment Report (AR6) of the Intergovernmental Panel on Climate Change (IPCC 2023) identifies climate change-related impacts and the consequent loss of biodiversity as the two main challenges with which the "global risk society" (Beck 2011) will have to contend, both because of the growing impact on cities and territories and the empirical evidence of the economic, social, and environmental damage that results (Latour 2017).

The ecological and digital transition, adaptation to climate change, the promotion of sustainability and efficiency of cities from a Smart perspective, and strategies aimed at ensuring the protection, restoration and enhancement of ecological-environmental components, have, therefore, become pivotal objectives that EU member states, including Italy, have decided to pursue, as they constitute a key junction for relaunching a smarter Europe, through innovative economic transformations; greener and low-carbon; more connected, through regional mobility and connectivity; more social, through implementation of the European pillar of social rights; and closer to citizens, through sustainable and integrated development of urban, rural and coastal areas through local initiatives (Giffinger 2007).

In this way, the sense of the guidelines enshrined by the United Nations with the identification of the 17 "Sustainable Development Goals" (UN 2015), within the "2030 Agenda for Sustainable Development" (2015), adopted by member states in 2015, and, in particular, the need to "make cities and human settlements inclusive, safe, resilient and sustainable" (SDG 11), through experimenting with forms of innovation for the improvement of citizens' living conditions and for the cultural, economic and social growth of communities, consistent with the goal of "protecting, restoring and promoting the sustainable use of terrestrial ecosystems, sustainably managing forests, combating desertification, halting land wastage and halting biodiversity loss" (SDG 15), including through the adoption of "urgent measures to combat climate change and its impacts" (SDG 13).

More recently, these themes have found further operational references and specific funding channels in the European Green New Deal (EU 2019), the EU's main growth strategy aimed at transitioning to a sustainable economic model that will achieve the goal of making Europe the first climate-neutral continent by 2050; in the Just Transition Fund (EU 2021), aimed at providing support to territories facing serious socio-economic challenges resulting from the transition to climate neutrality, and in the new EU programming 2021/2027 Horizon Europe (EU 2020), which introduces, as a core principle, the need for change in the EU economy and society to reduce environmental degradation and risk, halt and reverse biodiversity decline, and sustainably manage key resources while meeting the EU's established climate targets.

Specifically, the *Food, Bioeconomy, Natural Resources, Agriculture and Environment Cluster* aims to ensure a balance between environmental, social and economic objectives through actions aimed at: "restoring ecosystems and biodiversity; sustainably managing natural resources to ensure food security and a clean and healthy environment; making Europe the first circular, climate-neutral and sustainable economy through the transformation of mobility, energy, construction and production systems; and creating a more resilient, inclusive and democratic European society, prepared for and responsive to threats and disasters, addressing inequalities and providing high-quality health care, empowering all citizens to be able to undertake green and digital transitions." Goals shared with the Climate, Energy and Mobility Cluster as part of the European Commission's long-term strategic vision "*A Clean planet for all*" (EC 2018) for "a prosperous, modern, competitive and climate-neutral economy", emphasizing the interdependence between the transition to a neutral economy and the energy and mobility sectors.

Finally, numerous scientific researches have highlighted the improvement of human health and well-being as a priority urban goal and have increasingly promoted the creation of safe, healthy, inclusive, and sustainable urban environments in cities, emphasizing, in particular, the importance of building networks of green spaces and public spaces as places for gathering and relationships that stimulate social cohesion, reduce crime rates, promote the use of sustainable mobility and pollution reduction, and contribute to the pursuit of urban welfare (WHO 2016; UN 2015).

5.2 Green Connective Networks for the New Urban Welfare

In this new context of reference, there is, therefore, the need to respond to the demands of the "new urban question" (Castells 1979; Secchi 2011), which is characterized by a profound crisis in the environment, social relations and mobility, further aggravated by the consequences induced by the pandemic, and which has called for a renewed disciplinary attention and an innovative and integrated approach in the search for new paradigms for a sustainable transformation of the territory, in the dual meaning of "ontological foundation and operational orientation" in the processes of "reterritorialization" activated by a renewed "consciousness of place"(Magnaghi 2020).

New paradigms aimed at the construction of a new urban *welfare*, which assumes the public city as the reference matrix "of a tangible and intangible, interactive and integrated 'network of networks' that guarantees a capillary territorial garrison and endowment" (Ricci 2020a, b).

In this framework, the connective networks of green and the networks of natural and anthropogenic components structuring the territory, constitute a "transversal reticular paradigm" (Donadieu 2006; Gambino 2010) of the planning process, characterized by a multi-scalar and multi-sectoral approach, able to enhance, within a broader regeneration strategy, the interactions between the different connective and network systems that cross the territory (the hydrographic network, the ecological network, the cultural heritage network, the agricultural fabric network and the soft mobility network) (Ricci et al. 2019).

The EU definition (EC 2013) on Green Infrastructures recalls three important aspects; the concept of ecological connectivity, understood as a network of systems and/or land areas, environments and landscapes; the key role of planning; and the concept of multifunctionality of ecosystems (Hansen and Pauleit 2014) which, as outlined by the European Commission in 2012 (EC, 2012), refers to the wide range of functions that GIs can provide-including protecting ecosystems and biodiversity, improving the functionality of ecosystem services, promoting the health and well-being of society, and supporting the green economy-in order to ensure the essential conditions for the sustainability of transformation processes (EU 2020).

This system of networks assumes a relevant strategic role in the activation of ecological connections within complex metropolitan urban contexts (Arcidiacono et al. 2016) and in the "search for co-evolutionary pathways of economic and ecological systems and thus innovative forms of interaction between environment and society" (Gambino 2004), contributing to the reduction of the fragmentation of natural habitats and the ecological footprint of urbanized environments, as well as to the increase in the degree of biological diversity and the degree of resilience of the urban ecosystem (Voghera and Giudice 2019). Contributing, moreover, to the experimentation of urban regeneration strategies and interventions, at local and metropolitan scales, that place at the centre the profound link between the enhancement of historical-cultural identity, quality and sustainability of urban contexts and socioeconomic development opportunities, as also reiterated in the *European Framework for Action for Cultural Heritage* (EC 2019), which promotes integrated and participatory approaches to produce positive impacts on the four pillars of sustainable development, economy, culture, society and environment.

In this context, natural and cultural heritage is configured as a strategic engine of cultural-led *urban regeneration* (Paddison and Miles 2006), through the adaptive reuse of heritage that configures a circular process of multidimensional value production for the regeneration of marginal parts of cities. In this way, natural and cultural heritage begins as a resource to stimulate development, innovation and activation of new economies aimed at culture, tourism, and environmental sustainability, to improve citizens' quality of life.

> Negli spazi dilatati della città diffusa, è l'intero apparato paesistico della campagna acculturata e degli stessi spazi naturali, e non più soltanto l'insieme pur ricco dei luoghi centrali, a essere coinvolto nei nuovi processi di significazione, di comunicazione sociale e di organizzazione urbana. E' il territorio storico nella sua globalità, con le sue reti organizzative, le sue geometrie latenti, i suoi palinsesti paesistici e il suo articolato patrimonio culturale, a costituire il teatro dei processi urbani contemporanei (Gambino, 2004).

The contribution is contextualized in the framework of the author's research activities on the topic of the metropolitan dimension of the contemporary city and government models for large-area spatial planning, concerning experiences where experiments in inter-municipal associationism are underway. Among the research, in particular, the chapter refers to Sapienza Research Project, University of Rome, Department of Planning, Design, Architectural Technology, "*Metropolitan dimension and inter-municipal planning. Criteria (multiple) of cooperation perimeter and development perspectives of Lazio territories in a European comparison*" (Scientific Responsible Prof. Carmela Mariano).

In particular, the contribution aims to present one of the case studies investigated in the research on the topic of the inter-municipal dimension as a minimum territorial reference for the exercise of the complex functions of environmental and landscape protection and enhancement, through an approach that substantiates, also methodologically and operationally, the adoption of an inclusive conception of territory for the purpose of a unified knowledge and design of the different components.

A conception that "assumes the environment and the large-scale relationships related to it, as the matrix of reference for any assessment of possible transformations, making itself the bearer of environmental content, concerning ecological aspects both for natural cycles and for the effects of anthropic interventions on them (natural/anthropic); that considers the organization of space and the landscape as the interpretive key and term of reference for design inputs" (Ricci 2005).

In this framework and, in particular, in the described experience of the Scheme of Spatial Coherence (SCOT) of the Grenoble Urban Region, the system of networks between natural, semi-natural and human components represents the framework for the implementation of urban regeneration strategies and metropolitan territorial rebalancing declined in the PLUi (Plans Locaux d'Urbanisme intercommunaux) and, at the same time, the invariant for the development and reconfiguration of the physical and socio-economic assets of the vast area territory, using the concepts of nature and landscape as catalysts for regeneration and urban *welfare* (Desjardins and Leroux 2007).

The Scheme of Spatial Coherence (SCoT) is a strategic planning tool extended to the territorial sphere of a vast area, which makes it possible to coordinate, in a logic of sustainable development, the various policies of territorial government in terms of spatial and urban planning, housing, economic activity, mobility or environment. In view of the specificities and transformation dynamics of its territory, it identifies strategies and objectives aimed at management, planning and development, which thus constitute a framework for the fielding of various public policies.

The case study is set in the French context, which has been profoundly innovated by the institutional reform of L. 58/2014 (MATPAM) "*Loi de modernisation de l'action publique territoriale et d'affirmation des métropoles*" and the previous *Grenelle* Laws (2009/2010) that had initiated a progressive consolidation of national environmental policies, within a profound process of renewal of the planning system at all levels (Ruffolo 2009).

A profound overhaul of the institutional architecture that has resulted in a shift from inter-municipal government structures, characterized by the exercise of a single competence, to structures with multiple vocations, which has triggered a gradual transformation of "management" inter municipality, used in the first phase to ensure administrative efficiency and proper service delivery, to "project" inter-municipality (Deffigier 2007), which presupposes a higher level of cooperation and implies a more advanced and structured planning and coordination process (Mariano and Valorani 2018).

5.3 The Vision of the SCoT of the Grenoble Urban Region

At the regional level, in France, the Schéma Régional de Coherence Écologique (SRCE) provides the strategic reference scheme of "Trame verte et bleue" for the implementation of planning instruments at the local level (Plan Local d'Urbanisme and Plans Locaux d'Urbanisme intercommunaux) and inter-municipal wide area (Schéma de la Coherence Territoriale) (Poli and Ravagnan 2017).

The case of the (SCoT) Schéma de Coherence Territoriale de la Région Urbain de Grenoble (approved in 2012 and updated in 2018), a territory in the Rhône-Alpes et d'Auvergne region of more than 3746 km^2 in area (including 3/4 natural, agricultural, or woodland areas), consisting of 7 territorial areas(établissement public de coopération intercommunale EPCI) characterized by well-defined territorial identities, 268 municipalities and a population of about 800,000 inhabitants, is set in this reference context.

Situated between the Belledonne, Vercors and Chartreuse massifs, the Grenoble region covers a diversified relief between plains and mountains and boasts a rich heritage in terms of natural spaces and biodiversity. The agglomeration's natural wealth is based on the diversity of natural environments, the dynamism of agricultural areas and valleys that have a high agronomic value, now partially compromised by the rapid expansion of urbanization (Figs. 5.1 and 5.2).

The strategic vision of SCoT is structured around some structuring choices:

- Preserve and sustainably develop natural resources and the green and blue network as a condition for the development of agriculture and forestry.
- Improve the quality of the living environment by integrating environmental, landscape, safety, and health requirements into land use planning.
- Strengthen the attractiveness of the metropolitan area by meeting the challenges of sustainable development.
- Balancing and polarizing regional development to counter the suburbanization and dispersion of urban functions.
- Intensify spatial planning and strengthen the mix of functions to combat urban sprawl and land consumption.

This vision is organized around three macro-strategies, "Preserver, Èquilibrer, Organizer."

Preserver: save/increase space and resources.

The space and environment that contribute to the attractiveness and overall quality of the region are common goods that must be managed, optimized, and enhanced for the benefit of all.

Thus, the goal is to organize the Grenoble region from the protection of natural, aquatic, agricultural and forest areas, in view of their role in structuring the region in ecological, social, landscape, heritage, economic, security and public health terms.

Èquilibrer: distribute and quantify the capacity of each sector without reducing agricultural and natural areas.

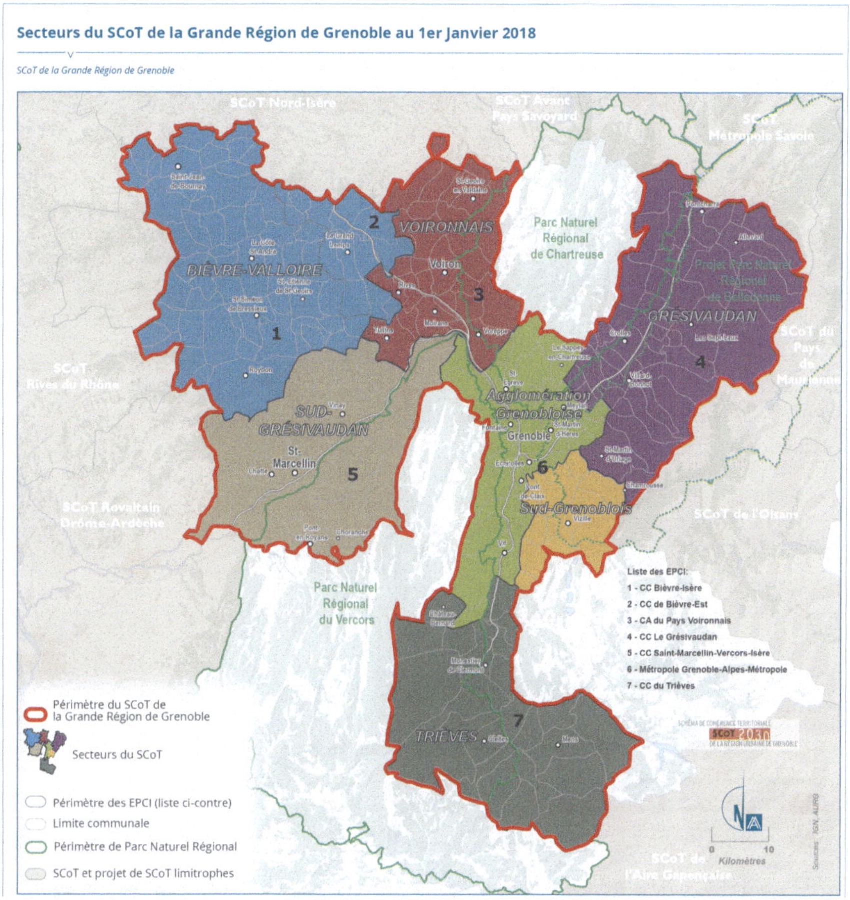

Fig. 5.1 The spatial structure of the Scot of the Grenoble Urban Region. Schéma de Cohérence Territoriale de la Région Urbaine de Grenoble (2012 update 2018), Agence d'urbanisme de la région grenobloise

It expresses the plan for a concerted, more equitable, more united, and more functional spatial organization that revitalizes small towns and rural villages and prioritizes residents' living comfort based on "short distances."

Organizer: Organizing the region around living spaces of higher quality and density.

The SCoT has designed a network of attractive and complementary urban and rural centres to make the daily lives of residents easier, more convenient, more pleasant, and more supportive within a more qualitative and less space-consuming shared living space: a diverse and quality supply of housing, services, and facilities; proximity to public transportation; a space for nature in the city, etc.

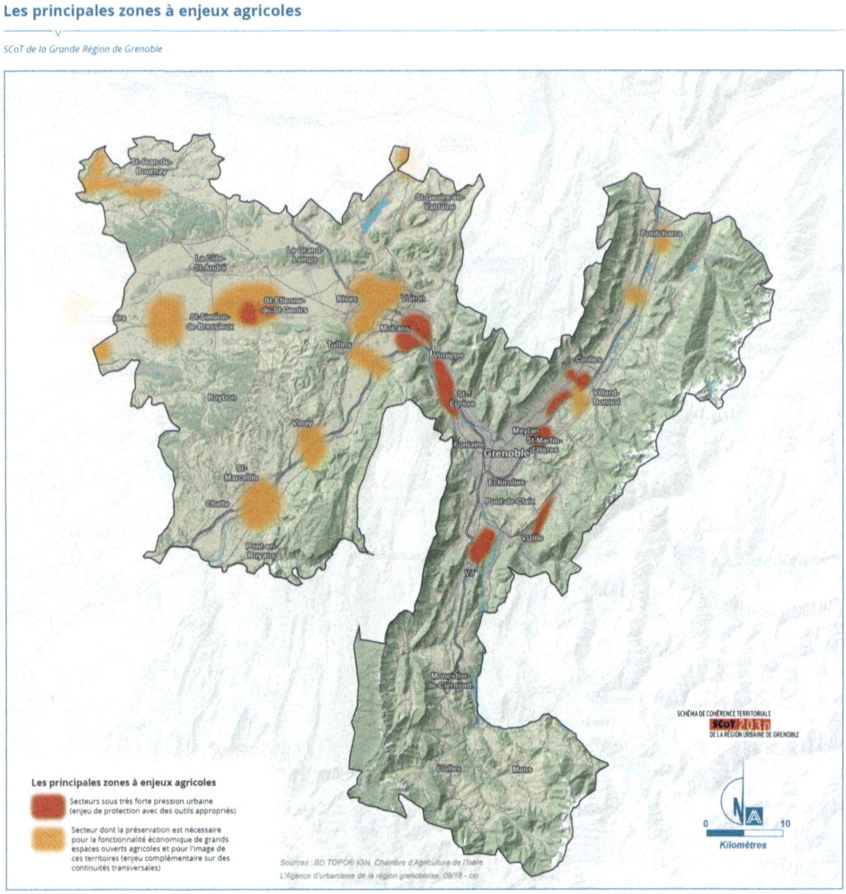

Fig. 5.2 The main areas with agricultural function in the Scot of the Grenoble Urban Region. Schéma de Cohérence Territoriale de la Région Urbaine de Grenoble (2012 update 2018), Agence d'urbanisme de la région grenobloise

5.4 Strategies for Ecological and Environmental Regeneration

The Grenoble region is committed to a general policy of conserving the territory's natural resources. Figure 5.3 identifies natural continuities that are important for the balance of the territory and its ecological functionality in consistency with neighbouring areas, highlighting the goal of preserving large natural mountain environments, forest complexes, agro-pastoral intermediate zones, and predominantly agricultural areas. In particular, the preservation of alluvial forests and wetlands, so important for ecosystem functioning, is highlighted.

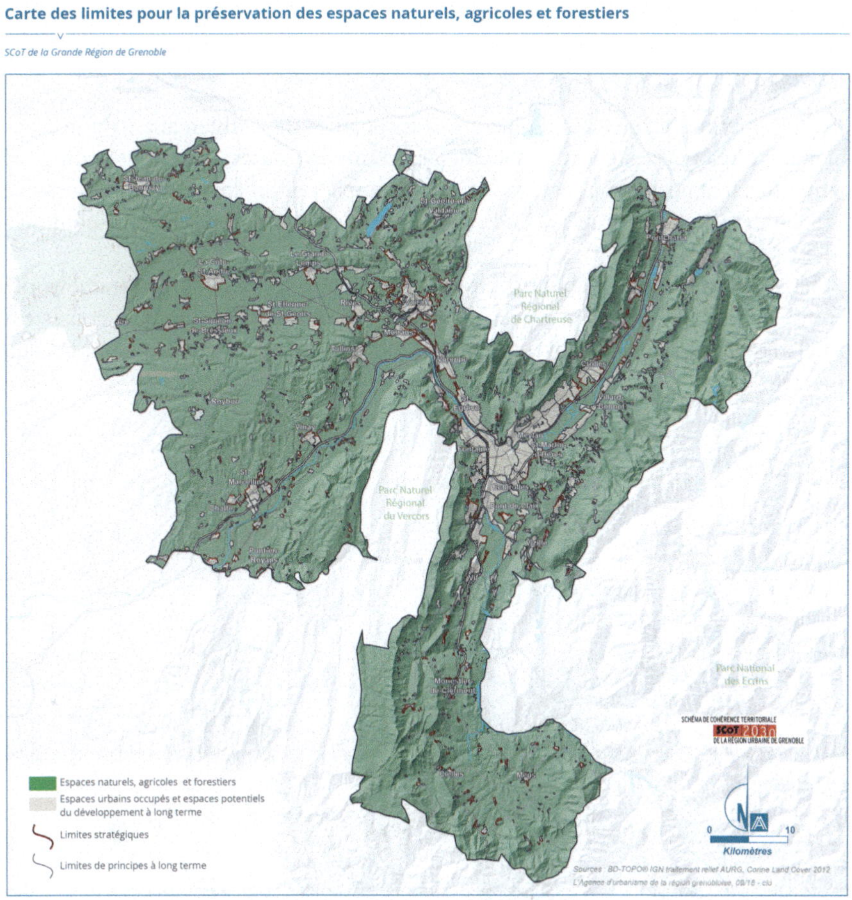

Fig. 5.3 Boundary map for the protection of natural, agricultural and forest spaces. Schéma de Cohérence Territoriale de la Région Urbaine de Grenoble (2012 update 2018), Agence d'urbanisme de la région grenobloise

The map also highlights the important natural terrestrial and aquatic connections that protect the ecological functionality of the area in interaction with the natural environments bordering the Grenoble region (Chartreuse, Vercors, Belledonne, Oisans in particular).

From the recognition of these invariants, the SCOT identifies guidelines and goals for urban development, identifying:

- "strategic boundaries" of agricultural, natural and forest areas (in green), which are to be protected from urbanization and which, once defined in local planning instruments, become permanent.
- "principal boundaries," the delimitation of which by local urban areas may change over time, depending on the potential development of certain territories.

– the "potential areas for long-term development" within which local planning tools will be able to define the conditions and rules for compatible urban development.

The SCoT defines a "green and blue grid." It specifies the emblematic natural areas which are already the subject of specific protection measures, as well as ecological continuities that need to be preserved from the impact of urbanization and excessive fragmentation by infrastructure and restored to good condition where discontinuities exist (Fig. 5.4).

This network should enable the Grenoble region to create a "green framework" of reference within which to place a development project for the area, consistent with its natural spaces, to prevent possible conflicts between settled functions and protected areas.

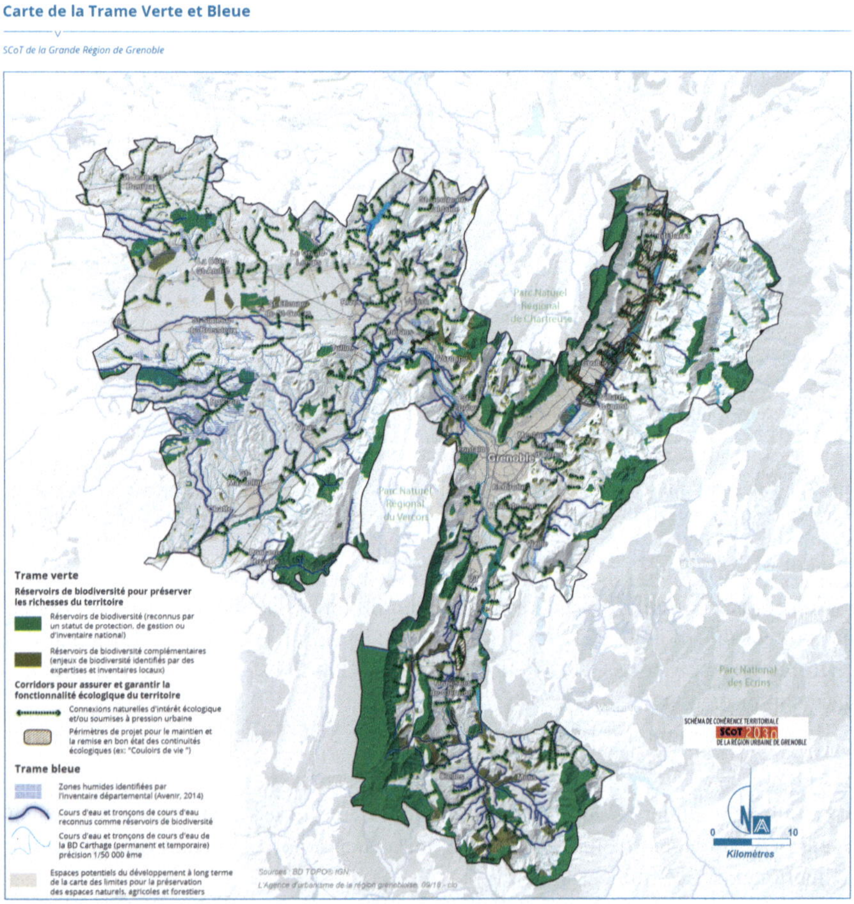

Fig. 5.4 Green and blue plot map. Schéma de Cohérence Territoriale de la Région Urbaine de Grenoble (2012 update 2018), Agence d'urbanisme de la région grenobloise

Specifically, the goals are:

- to conserve natural resources, particularly reservoirs of biodiversity, terrestrial and aquatic areas of high ecological quality, which will be preserved and cannot be endangered. The envelopes proposed in the SCoT will, if necessary, be modified and specified at the municipal level in the Plan Local d'Urbanisme (PLU) (Fig. 5.5).
- to ensure the ecological functions of the region by preserving floodplain forests and wetlands, the fragile environments recognized for their functional role; important natural continuities (forests, wetlands, etc.) and agricultural areas; wetlands, whose functional role and biodiversity must be protected from sources of pollution; ecological corridors, natural connections, and barriers to urbanization. They include natural, terrestrial, aquatic, semi-natural and agricultural areas, areas of discontinuity to be restored to repair the damaging effects of fragmentation on the continuity of flora and fauna, caused by the caesuras brought about by urban development, especially infrastructure.

The quality, diversity and coherence of natural and urban landscapes and townscapes are among the key values underlying a region's attractiveness. These structuring elements are given renewed attention in Grenoble's land development vision, which defines goals for the enhancement of landscape structuring elements (Fig. 5.6). These include:

- Preserve the region's major iconic sites from uncontrolled urban pressures.
- Enhance the visual fronts of forests, vectors of identity and background of many landscapes in the region.
- Preserve and enhance the landscape qualities of the area and, more generally, all the elements that contribute to mountain tourism.
- Enhance the road system as a means of landscape discovery and enhancement, particularly by developing thematic tourist routes.
- Preserve and improve the quality of "entrances to cities" near major traffic arteries.
- Improve the quality of areas around transit routes and strengthen links with rivers, banks, and roads.

Figure 5.7 qualifies the scenic interest of the roads and the type of transit. These routes (historic roads, steep-slope roads, scenic roads, access to mountain ranges, tourist routes) play an important role in how the region is perceived.

The guidelines provided in the SCoT associated with this map can allow for the avoidance of linear urbanization along major roads to improve the quality of the surrounding environment and preserve views of the broader landscape from highways and major thoroughfares (ensuring the continuity of the highway network), redeveloping major inter-municipal links, major urban arteries, boulevards, and urban crossings; and preserving views of the broader landscape from scenic and tourist routes.

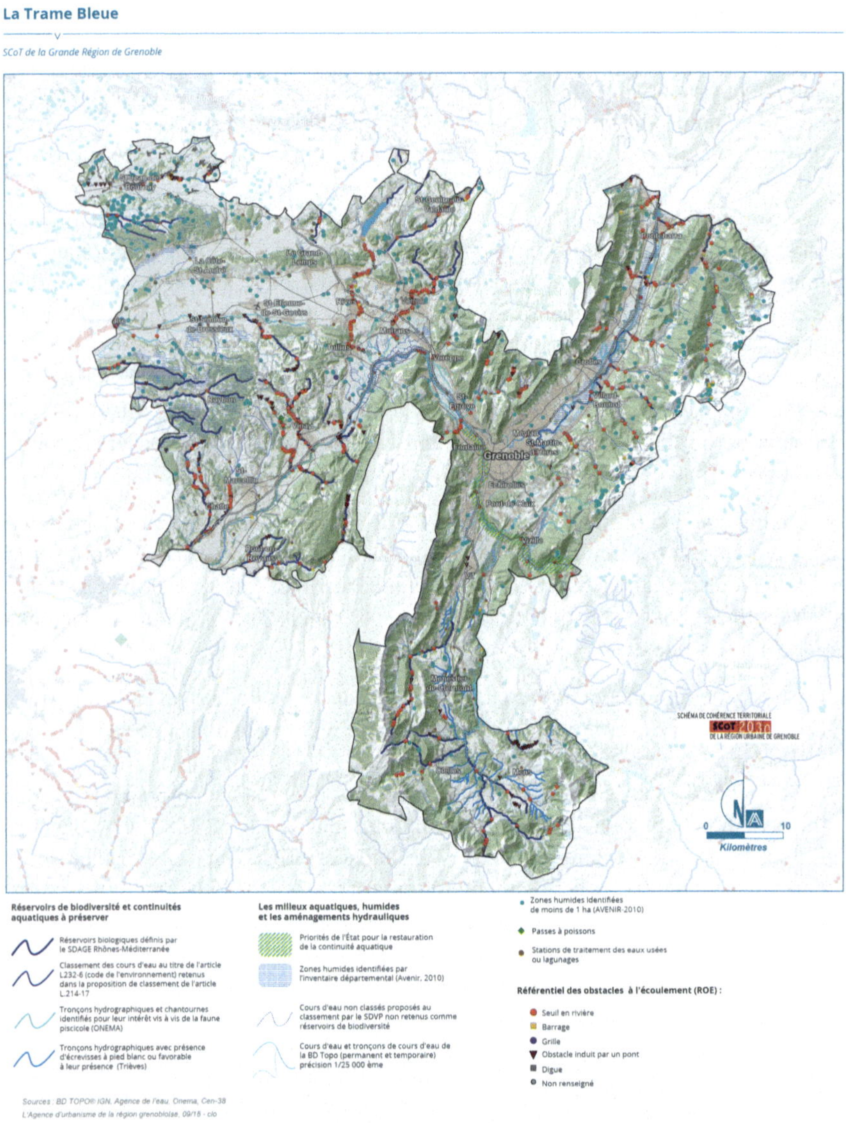

Fig. 5.5 Blue plot map. Schéma de Cohérence Territoriale de la Région Urbaine de Grenoble (2012 update 2018), Agence d'urbanisme de la région grenobloise

5.5 Conclusion

The analysis of the case study and, more generally, the research carried out on various European case studies, is consistent with the thematic focus of the volume *Rural Urban Dialogue in Green City Design*, regarding the role of urban green systems,

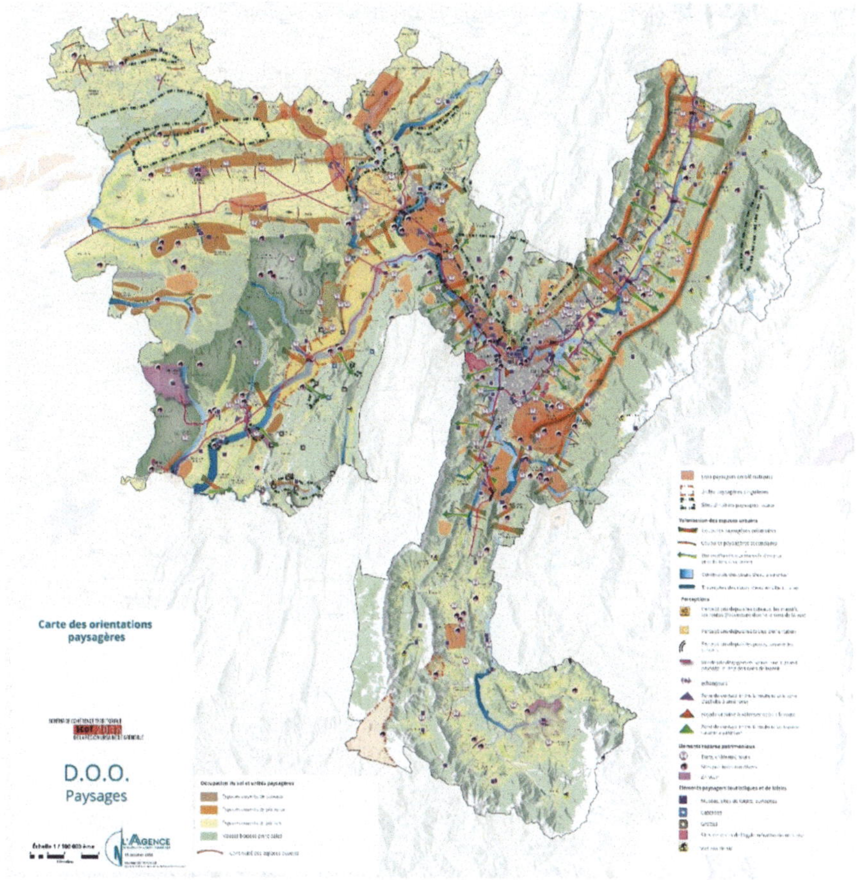

Fig. 5.6 Map of land occupancy and landscape units. Schéma de Cohérence Territoriale de la Région Urbaine de Grenoble (2012 update 2018), Agence d'urbanisme de la région grenobloise

landscape settings and ecological-environmental networks in territorial regeneration strategies capable of improving the urban environment, health, wellbeing, and quality of life.

The new dimension of the contemporary city and the need to share structural choices of policy and sustainable development at a supra-municipal scale calls for the need to address, in a unified and integrated way, issues common to the territories of vast areas, to ensure their proper functioning (Territorial Agenda of the European Union 2020; EU 2011).

Numerous scientific researches have also highlighted the improvement of health and human well-being as a priority urban objective and have increasingly promoted the creation of safe, healthy, inclusive and sustainable urban environments in cities, emphasising, in particular, the importance of building networks of green spaces and public spaces as places of aggregation and relations that stimulate social cohesion,

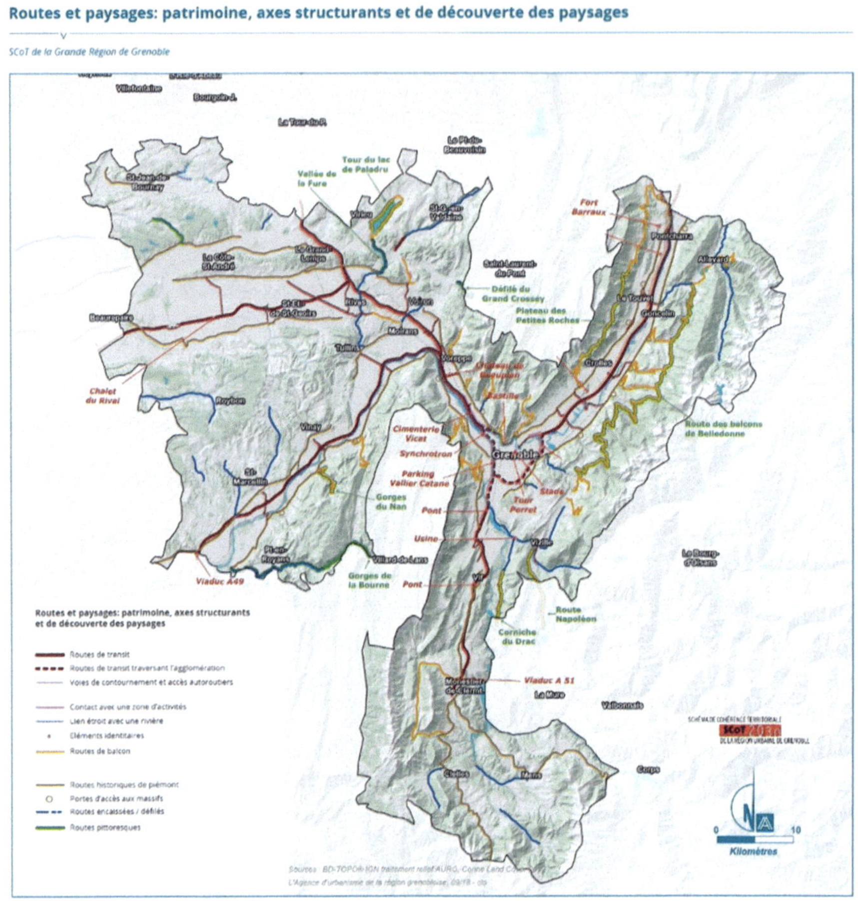

Fig. 5.7 Map of routes and landscapes. Schéma de Cohérence Territoriale de la Région Urbaine de Grenoble (2012 update 2018), Agence d'urbanisme de la région grenobloise

reduce crime rates, promote the use of sustainable mobility and the reduction of pollution and contribute to the pursuit of urban welfare (WHO 2016; UN 2015).

The best practices analysed in the research highlight the use of practices and tools that recognise landscape as the guiding element for addressing the strategies underlying territorial development policies, acknowledging the "valenze propositive insite nella nozione di paesaggio – della sua capacità di porre le premesse, attraverso l'individuazione e la descrizione delle identità paesistiche, per una salvaguardia e valorizzazione dei connotati costitutivi dei luoghi in cui si addensano valori di natura e valori di memoria, di mostrare e suggerire modi per l'organizzazione dello spazio, per la sua abitabilità e fruibilità da parte degli uomini" (Gambi 2008).

Consistent with this conception, the framework of planning tools can be innovated by assuming the environment, the landscape and the large-scale relations connected to it, as key factors in urban regeneration processes that actively contribute to the promotion of inclusion and wellbeing as well as sustainable economic development.

References

IPCC (2023) Climate change 2023 synthesis report, available to IPCC_AR6_SYR_LongerReport.pdf
Latour B (2017) Down to earth—politics in the new climatic regime. Polity Press, Cambridge
Beck U (2013) La società del rischio. Verso una seconda modernità, Carocci editore
Giffinger R & al (2007) Smart cities: ranking of European medium-sized cities. Wien: Centre of Regional Science (SRF) Wien University of Technology
UN (2015) 2030 agenda for sustainable development. UN, New York
EU (2019) European green new deal, communication from the Commission to the European Parliament, the Council, the European Economic and Social Committee and the Committee of the Regions, Brussels, 11.12.2019
EU (2021) Regulation 2021/1056 of the European Parliament and of the Council of 24.6.2021 establishing the Just Transition Fund
EC (2021) Investing to shape our future. Horizon EU Programme 2021/2027. Webpages: https://ec.europa.eu/info/sites/info/files/research_and_innovation/funding/presentations/ec_rtd_he-investing-to-shape-our-future.pdf
EC (2018) A Clean Planet for all. A European strategic long-term vision for a prosperous, modern, competitive and climate neutral economy, Communication from the Commission to the European Parliament, the European Council, the Council, The European Economic and Social Committee, The Committee of The Regions and the European Investment Bank, Brussels, 28.11.2018, available to https://eur-lex.europa.eu/legal-content/EN/TXT/?uri=CELEX%3A52018DC0773
WHO (2016) World Health Statistics 2016: Monitoring Health for the SDGs, Sustainable Development Goals; World Health Organization: Geneva, Switzerland, 2016; Available online: https://apps.who.int/iris/handle/10665/206498
Castells M (1979) The urban question a Marxist approach. MIT Press, Cambridge
Secchi B (2011) La nuova questione urbana: Ambiente, mobilità e disuguaglianze sociali. Crios 1:89–99. https://doi.org/10.7373/70210
Magnaghi A (2020) Il principio territoriale. Bollati Boringhieri, Torino
Ricci L (2020) Città contemporanea e nuovo welfare. Una rete di reti per rigenerare la città esistente, in Città esistente e rigenerazione urbana. Per una integrazione tra urbs e civitas, by Poli I Aracne editrice, Roma
Ricci L, Iacomoni A, Mariano C (2020) Regole, strumenti, meccanismi e risorse per una politica integrata di produzione di servizi, in Ananke n. 90
Gambino R (2010) Lectio Magistralis Parchi e paesaggi d'Europa. Un programma di ricerca territoriale, in Ri-Vista. Ricerche per la progettazione del paesaggio, n. 14, 3–20. Available online: https://issuu.com/dida-unifi/docs/14r_2010
Ricci L, Mariano C, Poli I (2019) Green infrastructures in the masterplan of Rome. Strategic components for an integrated urban strategy. In: Gargiulo C, Zoppi C (eds) Planning, nature and ecosystem services. FedOAPress - Federico II Open Access University Press
EC (2013) Green infrastructure (GI)—enhancing Europe's natural capital, Communication from the Commission to the European Parliament, the Council, the European Economic and Social Committee and the Committee of the Regions, Brussels, 6.5.2013
Hansen R (2014) Pauleit S (2014) from multifunctionality to multiple ecosystem services? A conceptual framework for multifunctionality in green infrastructure planning for urban areas. Ambio 43:516–529. https://doi.org/10.1007/s13280-014-0510-2

EC (2012) The multifunctionality of green infrastructure. http://ec.europa.eu/environment/nature/ecosystems/docs/Green_Infrastructure.pdf
EU (2020) Biodiversity strategy for 2030.Available online: https://environment.ec.europa.eu/strategy/biodiversity-strategy-2030_en
Gambino R (2004) Conservare, innovare. Paesaggio, ambiente, territorio, UTET Libreria
Arcidiacono A, Ronchi S, Salata S (2016) Managing multiple ecosystem services for landscape conservation: a green infrastructure in Lombardy Region. Proc Eng 161:2297–2303
Voghera A (2019) Giudice B evaluating and planning green infrastructure: a strategic perspective for sustainability and resilience. Sustain For 11:2726
EC (2019) Direzione generale dell'Istruzione, della gioventù, dello sport e della cultura, Quadro d'azione europeo sul patrimonio culturale, Ufficio delle pubblicazioni, https://data.europa.eu/doi/10.2766/622226
Paddison R, Miles S (2006) Cultural-led urban regeneration, Routledge
Ricci L (2005) Diffusione insediativa territorio e paesaggio. Un progetto per il governo delle trasformazioni territoriali contemporanee, Carocci Editore, Roma
Desjardins X, Leroux B (2007) Les schémas de cohérence territoriale: des recettes du développement durable au bricolage territorial. In FLUX 2007/3, n. 69, 6–20
Ruffolo G (2009) La nuova politica ambientale francese, in Gazzetta Ambiente, n. 4, 5–9
Deffigier C (2007) Intercommunalitè et territorialisation de l'action publique en Europe, Revue française d'administration publique 2007/1, n. 121–122, 79–98
Mariano C Valorani C (2018) Territori metropolitani e pianificazione intercomunale, FrancoAngeli editore
Poli I, Ravagnan C (2017), Trame verdi e blu: verso un futuro affidabile tra visione strategica e gestione dei rischi/Green and blue networks: towards a safe future within risk management and strategic vision, in Urbanistica, n. 160, INU edizioni
Donadieu P (2006) Campagne urbane. Una nuova proposta di paesaggio della città, ed. italiana a cura di M.V. Mininni, Donzelli editore
Schéma de Cohérence Territoriale de la Région Urbaine de Grenoble (2012a), Comite syndical de la région grenobloise, Agence d'urbanisme de la région grenobloise, Le Rapport de présentation, 21 december 2012, update 2018
Schéma de Cohérence Territoriale de la Région Urbaine de Grenoble (2012b), Comite syndical de la région grenobloise, Agence d'urbanisme de la région grenobloise, Le projet d'aménagement et de développement durable (PADD), 21 december 2012, update 2018
Schéma de Cohérence Territoriale de la Région Urbaine de Grenoble (2012c), Comite syndical de la région grenobloise, Agence d'urbanisme de la région grenobloise, Le document d'orientation et d'objectifs (DOO), 21 december 2012, update 2018
EU (2011) Territorial Agenda of the European Union 2020, Towards an Inclusive, Smart and Sustainable Europe of Diverse Regions, available online: territorial_agenda_2020.pdf (europa.eu)
Gambi L (2008) La cognizione del paesaggio. In: Guermandi MP, Tonet G (eds) Scritti di Lucio Gambi sull'Emilia Romagna e dintorni, Bologna

Chapter 6
Cultivating Public Spaces—Method and Procedures for Inclusive Urban Re-appropriation Practices

Francesca Rossi and Margherita Meta

Abstract This chapter delves into the transformative potential of cultivating public green spaces within economically distressed and often marginalised public settlements. These areas, mainly inhabited by poor citizens, offer a unique opportunity to promote social cohesion, support the circular economy and encourage solidarity-based citizenship practices. Informal cultivation initiatives typically take place in green spaces of various sizes, often municipally owned but managed by private citizens, which organise themselves in specific associations which employ physical revitalisation strategies to reclaim neglected public green spaces within their settlements. Although without any professional agricultural experience, citizens grow vegetables for the immediate sustenance of their family, securing these spaces through concessions from the municipality, primarily aiming to meet the settled community's needs with sustainable produce. Beyond mere economic support, these informal cultivation areas act as crucial social glue, compensating for the absence of gathering spaces and promoting a conscious use of resources. They nurture a sense of community, encourage responsible use of public spaces, facilitate active citizenship, and foster peaceful interactions among residents. They also implement adaptive solutions, circular economy practices and innovative cultivation techniques, supporting a healthier and resilient lifestyle, challenging consumerism, and the uncontrolled urbanisation of green areas.

Keywords Urban gardens · Social cohesion · Appropriation practices · Sustainability · Circular economy

F. Rossi (✉) · M. Meta
Department of Planning, Design, Technology of Architecture (PDTA), Faculty of Architecture, Sapienza University of Rome, Rome, Italy
e-mail: fra.rossi@uniroma1.it; margherita.meta@uniroma1.it

A. Battisti et al. (eds.), *Urban-Rural Dialogue in Green City Design*, Designing Environments, https://doi.org/10.1007/978-3-031-88318-7_6

6.1 Urban Gardens—Conceptual Definition and Contextualisation in Italy

Agricultural practice in Italy has progressively shifted to the sector defined as "tertiary agriculture," becoming part of that economic sector in which "goods are produced or services are provided, i.e. all activities that are complementary and supportive to those of the primary (agriculture, livestock, etc.) and secondary (manufacturing) sectors, in short, intangible services that may or may not be incorporated into a good" (Ricci 2007).

Tertiary agriculture constitutes a sectorization of agricultural practice within the so-called "advanced tertiary system," in which the production resulting from the work derived from agricultural practice is no longer exclusively classified in the economic category of the primary sector, but it also encompasses the classification of production processes different from others, in which "semantic skills and symbolic manipulation define the production or service chain in which this process is inserted" (Reich 1991). This happens because the practice of cultivating public spaces moves into the cluster of practices that intersect with the typical issues of the third sector and civil economy, characterised by the effort of integrating strategies of social responsibility and support within it.

Such a premise is essential for understanding the circumstances resulting from the cultivation of public space, in terms of transforming territorial socio-economic systems, communal models, and, globally, the various forms of inhabiting the territory. This understanding holds significant utility in the aim of establishing regulatory models suitable for each context where it must be encompassed in a flexible yet organic standardisation procedure.

Regarding the Italian experience, there are a multitude of phenomena primarily oriented towards self-sustainability, localised in small parcels of land scattered in residual areas of rural zones. However, for the purposes of this article, the focus is on green spaces pertaining to condominiums or integrated into public spaces within the consolidated city. Contextualizing and subsequently regulating the practices of cultivating public space require a deep understanding of the dynamics that historically defined and regulated the management of territorial systems, to be implemented according to the innovative potential offered by the experiences seeking regulation. This process surpasses the bureaucratic complexity of systems that have often characterised the management of traditional social models related to contractual forms through which agricultural territories regulated their productivity, following a statist model that frequently stifled the possibility of new areas of specialisation and sectorization.

The system of "appropriation" by small social groups of cultivable land parcels finds its roots within the contractual models for the management of cultivable territories, which defined agricultural production relationships until the end of the last century and influenced the social patterns that formed around such cases (Pascale 2015).

This social system imparted a specific shape to the territories, giving most of the Italian landscape the form that still predominantly characterises it, marked by isolated cultivated land parcels in rural areas. Lands managed through the contractual form of sharecropping, a contract between the land labourer and the landowner, in which agricultural products were marketed within the markets of the nearest inhabited centres for the landowner's profit while simultaneously contributing to the sustenance of the labourer's family.

Otherwise, in the *latifondo* system, characterised by several small land parcels gathered around a multifunctional service centre known as "*masseria*." the peasants, living in the villages in the immediate vicinity, organised their work shifts under the employment of the landowner of the areas. Both contractual systems allowed for the self-sustainability of individuals working on the land, which, even before physically, was conceptually immersed in a microcosm of territorial and social systems that regulated its use and existence. These systems enabled the organisation of services necessary to fulfil productive functions, product valorisation for the market, and territorial care.

In the legal framework there exists no single and uniform regulation for the agricultural practice of social gardens in Italian legislation. In fact, the Civil Code only provides regulation for entrepreneurial activities related to the agricultural sector. The regulation of property on such lands is typically applied through the contractual form of "comodato d'uso" (loan for use), by local authorities if the land is publicly owned, or by the owners if the land is privately owned (Contratto di comodato d'uso, art. 1803 Codice civile).

The legal relationships established within the "comodato d'uso" contract regarding social agricultural practices are governed by provisions related to agricultural law. However, these provisions cannot extend to the regulation of aspects covered by business law, which does not regulate agricultural disciplines related to self-consumption.

Nevertheless, these practices of social cultivation for self-sustainability have crossed the boundaries of small rural communities and have also spread to small and medium-sized urban centres. Due to the diverse demand for services, new practices of social agriculture are emerging on land parcels, both public and private, that are unused and sometimes localised in marginal areas. The contractual relationship is established between the cultivator, often a citizen aiming for self-sustainability, who forms an association with others with similar interests, and the organiser. The organiser can be the legal entity that owns the land used or an intermediary who looks after their interests.

6.2 Urban Gardens as a Responsible and Collective Management Citizen Practice

The purpose of social agriculture practices in creating an innovative social service for the community, has positive global implications, including ecosystem and landscape aspects. These practices aim to promote social cohesion through the achievement of general-interest objectives. This includes promoting active citizenship through collective experiences of community farming, the physical and cultural reappropriation of territories where communities are settled, encouraging cultivation and waste disposal techniques that promote environmental sustainability, and, in general, engaging the public in a process of awareness and responsibility towards agricultural and environmental resources through principles related to the circular economy and distributive equity (Petrulli 2022).

Nevertheless, the legal regulation of the relationship between the grantor of cultivable areas and the usufructuary, causes an incredibly precarious nature of the regulation of such experiences. This precarious nature has implications on the ability to construct long-term conditions for experimenting with social agricultural practices that serve as a cohesive element for the community.

In fact, the organisation of "urban gardens" is inevitably linked to a process of exchange of goods and services that are resolved through an economic transition: whether the grantor of the areas for urban gardens is an agricultural enterprise, the same municipality in which these areas fall under, or a subject different from the municipality, the area concessionaires are required to contribute, bound to a specific timeframe in which the service will be guaranteed, but in no way does it certify the possibility of continuous use of the area if economic resources or the mere convenience at levels of investment in an immovable asset were to change. Many municipalities have issued regulations for social agriculture. However, there is no in-depth investigation into the possibility of collective management of real estate (the areas to be allocated), which very often remain unused and only become productive following a collective action that ensures the regeneration of the area through agricultural work and through good territorial practices which ensure social cohesion, the most valuable of gains, which should not be subject to privatisation, not even in the form of allocation to non-profit private associations (Pascale 2015).

The normative validation of experiences in social agriculture carried out on public or private property has implemented a paradigm shift, according to which urban gardens no longer play the generic role of reclaiming degraded areas to become the exclusive symbol of assistance for poor population segments. Instead, they assume the role of good citizenship practices that encompass a dual purpose: to prevent unused spaces from being abandoned, improperly used, or subject to vandalism, and on the other hand, to make these areas productive and equipped for public use, integrating both the landscape and social aspects (Pierri and Torquati 2016).

Indeed, an urban garden can be divided into one or more plots or minimum cultivation units, intended for individuals conducting non-professional productive activities, with multiple objectives such as family self-consumption, recreational use, educational purposes, and aggregative purposes among active groups sharing values of environmental sustainability and social inclusion. Therefore, the urban garden can be considered an expression of a more general process of modification of social needs" (Paci 2010).

The concrete difficulty in the creation of an innovative, public, and participatory system for managing the circumstances underlines the challenge of identifying a responsibility profile in the use of land for the creation of urban gardens. The absence of a single regulation allows local administrations to generally define their own autonomous rules for the assignment of green spaces, which include merit rankings aimed at guaranteeing priority access to individuals in social and economic distress (pensioners, unemployed individuals, low-income families, and so on). In this sense, the prevailing function remains indirect support for income resulting from the opportunity for personal and family self-consumption of the products obtained.

The main requirement is that cultivation and production activities must be non-professional, and that the commercialization of the harvested produce and the use of third-party individuals for paid land management operations are forbidden. In addition, organic cultivation methods are also preferred (Pascale 2015). Also the contractual forms for the allocation of land are not uniform: for public lands, the natural form is administrative concession; for lands owned by municipal administrations, long-term lease or loan for use is provided; in some particular cases, municipal administrations also foresee the assignment of areas to associations or non-profit constituted groups as an expression of a local community of citizens who will manage the spaces within (Pierri and Torquati 2016). Especially "educational gardens" can be realised and managed by professional agricultural entrepreneurs, in which the enterprise can simultaneously be the entity that carries out and manages the maintenance of the garden, providing a training service to users. Gardens with functions more oriented towards vegetables production for family consumption, the entrepreneur can ensure all technical services to guarantee the better use of individual cultivation units, such as wooden structures for equipment storage, internal roads, demarcation of plots, organic fertilisation, irrigation systems on request, and resting areas.

The analysis of the highlighted issues mainly suggests that the responsibility of all parties involved in the usufruct contract should be regulated, bilaterally, to counteract the precarious nature that affects it, especially in respect of the following aspects.

First, it would be necessary to specify the non-remunerative nature of agricultural production through standardised written agreements since the producer who markets agricultural products to third parties is criminally liable for damages, as they must guarantee the food safety of the products (Articles 114–115 of the Consumer Code), Second, it should be specified that the garden must be cultivated directly by the lessees identified in the agreement, and in the case of Associations or Established Groups, the names of the members should be communicated to the property to ensure access and use by eligible individuals, as well as specifying that any service provision can only be provided and requested by the contract holder (Pascale 2015).

In this way, the creation and management of urban gardens could better facilitate good citizenship practices in order to promote social cohesion, to create synergies between widespread local knowledge and established regulations with the aim of building essential knowledge frameworks for the development of local policies truly based on the needs and desires of citizens, to encourage civic activism and collective awareness regarding the need to experiment innovative care practices for the management and enhancement, of common goods and public spaces.

6.3 Urban Gardens as an Adaptive and Inclusive Strategy Facing Climate Change

Moreover, the characterisation of the collective function of urban gardens fits within the recent changes that the historical phase of the Covid-19 pandemic outbreak has brought about in widely heterogeneous contexts of contemporary society. As pointed out by David Harvey (1984), the crisis of territorial constraints generated by globalisation since the 1980s and 1990s has long been depriving populations of the support of territorial networks that allow societies to cohere around the concepts of territory, identity, and development. The pandemic crisis has, in some way, highlighted the resilience of territorial systems in response to the multiple challenges in the social and healthcare fields, wherein the search for areas for social agriculture activities to increase the resilience of cities in relation to their food supply capacities, has become dominant.

The basic principles of the "Urban Green Infrastructures" approach involve increasing the connectivity of green spaces, improving their capacity to provide various goods and services, integrating green infrastructure with other urban infrastructures, and ensuring collaborative and inclusive planning processes. Indeed, this type of information can contribute to increase the understanding of the contribution of environmental characteristics to human well-being and can help understand how social aspects influence the perception of the service, what kind of synergies or connections are created with other ecosystem services, and how experiences (individual or group) in places can influence urban development patterns (ibid). In fact, one of the most significant issues concerning the decision-making process faced by administrations in planning green infrastructures and managing urban green areas lies in the lack of precise and punctual information that goes beyond the classic spatial attributes and the ecological characteristics of the areas in question. The decision-making process of "urban green spaces"struggles to address the environmental and social challenges of areas designated for this purpose in an interdisciplinary and systemic manner.

Considering these preliminary observations, it is essential to emphasise the central role played by the implementation of sustainable and resilient projects for urban resilience spaces, such as urban gardens, given that cities are becoming more populous and are posing a challenge of increasing the number of consumers of agricultural products, in a scenario heavily destabilised by the pandemic crisis and leaning toward a decrease in supply.

From the environmental perspective. he central function of urban gardens is structured within long-established urban areas, starting with the metropolises in the Global South, where these self-sustaining cultivation spaces have taken on the role of providing food for vast populations from impoverished rural areas suffering from soil aridity, famines, and water shortages (Pettinati and Toldo 2018). Subsequently, urban agriculture practices have gained significance in the agricultural territories of large Western cities, where they have acquired socio-economic importance like other urban functions because they contribute to the surviving and well-being of the city. In these contexts, social agriculture goes beyond mere subsistence and takes on the role of interconnecting functions that contribute to the creation of adaptation mechanisms. The concept of an "urban garden"is now the result of a sustainable, community-based project, thanks to the synergistic interaction of the local public administration, on one hand, managing the allocation of plots and assessing the feasibility of project ideas, and citizen groups, on the other hand, submitting proposals for urban horticulture.

The resilience effects generated by the experience of urban community gardens and other forms of local agriculture are various and can be viewed from food, environmental, cultural, and social perspectives.

First and foremost, urban gardens represent a concrete source of food for communities, addressing the growing agricultural demand, especially in countries heavily reliant on foreign exports and facing the devastating effects of an inevitable decrease in agricultural production and the consequent rise in prices due to unforeseen circumstances. Moreover, they enable communities to adopt a sustainable production model that ensures variety and renewability of food resources: urban gardens fulfil the daily consumption of fresh vegetables expressed by every city, which is often challenging to achieve due to an economic model built on agricultural imports, in whose process, the emissions resulting from the transport of the goods contribute to exacerbating the harmful effects of the crisis triggered by the climate change.

From the inclusive perspective, the common purpose between the local municipal administration and citizen groups initiates an innovative, bottom-up social process that engages active citizen participation in the territory's development. Designing an urban garden means grappling with the idea of collective territorial construction in which every community member must participate, as those who inhabit and interact with the territory in various ways possess an in-depth understanding of its characteristics, needs, and potential.

The most significant expression of this collaborative dialectic that guides the planning of community gardens is represented by the choice of a short supply chain production system, which strengthens the network of interactions among geographically close groups of people who share the same local development project (Dansero et al. 2020).

Based on these considerations, it is inevitable to recognize the significant impact of urban gardens in terms of adaptive and collective participation in territorialisation processes.

6.4 Two Urban Regeneration Case Studies of Social Agricultural Practice

6.4.1 Orti Generali, Turin

The "Orti Generali" of Turin is a recovery case of an industrial dismissed area through a virtuous example of urban regeneration. A space dedicated to social agriculture practices in the southern part of the Mirafiori industrial area, where buildings grew rapidly to accommodate those working families who came to work at FIAT (the automotive industry), mainly from southern Italy.

This peripheral zone, clacking services, and social spaces, expresses the issues and contradictions typical of marginalised and degraded settlements in the outskirts that exploded in the 1980s. It was located within a river park, on the banks of the Sangone stream, urban gardens have emerged, open to citizens and visitors, thanks to a concession notice from the City of Turin.

Following the winning proposal for the physical and functional redevelopment of the area by the Municipality of Turin, the MIUR Smart Cities and Communities and Social Innovation competition selected the project presented by the Coefficiente Clorofilla association, now Orti Generali, to come to fruition.

After a research phase supervised by the Ca′ Foscari University of Venice from 2016 to 2019, the project materialised as a reality where citizens can "cultivate as a group, as a family, as an association, or individually; learn through theoretical and practical training courses, activities, and workshops on topics related to organic farming, horticulture, and rural work" (Orti generali Torino).

The early idea of the project, developed by Stefano Olivieri, Isabella Vecchi, and Matteo Baldo, the general coordinators of the association, started from an analysis of various possible actions in the area, which now undergoes redevelopment, integrated into a context where there were about 2000 unauthorised small and large gardens. In 2018, the city of Turin granted the management of the area and coordinated actions for the management of urban gardens around the formation of a "community pact" where gardeners take over the area for 15 years while caring for the land and maintaining a good level of decorum.

This pact has led to the participatory management of the space, which is now public and freely accessible (Milanesi 2023).

In the area of the "Orti Generali," there are 160 gardens of three different types—100, 65, and 50 m^2—with discounted rates for young people and those with limited means. To apply for garden management, it is sufficient to send an email to the municipality, and once granted, the concession lasts for 15 years because horticulture is seen as a continuous practice. Currently, the association is working to create another hundred gardens in the area facing the river, where remnants of unauthorised gardens effectively hinder access to water resources.

The project operates with attention to the water crisis affecting the Po Valley, planning the planting of sustainable crops that can also withstand potential extreme weather events resulting from climate change (Milanesi 2023).

The project's aim is to create a sustainable and inclusive environment where city dwellers can reconnect with nature by taking care of the space through gardening and volunteering. The project also collaborates with various NGOs and is part of neighbourhood solidarity networks.

In addition to this, the project strengthens the sense of belonging of the community by regenerating urban spaces through their collaborative participation. A care practice that also include attention to sustainability, in contrast to the disastrous effects of pollution and climate change: a specific land use organisation and specific material's quality levels were established during the project implementation (Vasile and Grasseni 2022).

The resulting landscape is a new and well-maintained environment, with new gardening infrastructures specifically imported from France called *ganivelles*, which were planted by volunteers to organise cultivation plots. A giving a sense of order and homogeneity, soon expanded by the installation of identical-looking sheds in all gardens. Gardeners, allowed to build greenhouses, were required to follow specific rules, and clearly differentiating the new gardens from spontaneous ones, enhancing the sense of care mentioned earlier, which acts as a community bond and spreads awareness of the importance of sustainability in human actions on the territories.

The "Orti Generali" also operate in accordance with the principles of environmental sustainability, allowing cultivation exclusively using organic methods. Ancient tree varieties are reintroduced in the area, with special attention to plants that promote the repopulation of pollinating insects. Technological innovation is embraced to serve the environment: a weather, humidity, and local temperature sensing control unit is employed to reduce water waste, activating a centralised irrigation system for all the gardens when needed. Social sustainability is achieved through the sponsorship of jobs and training internships, especially for poor individuals.

Finally, the "Orti Generali" project in Turin won the National Landscape Award 2022/2023 and aspires to increasingly transform into a participatory economy and even a community energy laboratory. In the motivations reported during the award ceremony, it was noted that "Orti Generali" was able to "recover the fabric of the pre-industrial agricultural landscape, collaborating with public entities, academic institutions, schools, associations, individual inhabitants" (Orti Generali Torino 2024).

6.4.2 Orti Urbani Garbatella, Rome

Garbatella is a neighbourhood located in the southern quadrant of the Metropolitan City of Rome, characterised by a settlement of public housing built in the 1920s. This settlement, built by the Autonomous Institute for Public Housing (IACP), distinguished itself with a low intensity building settlement system, aimed at evoking the bucolic atmospheres of the countryside and villages around Rome.

As the city centre was conceived to be a representative simulacrum of the grandeur of the imperial city, the neighbourhoods were designated for working-class settlement and meant to evoke simplicity and inspire visions of a simple and industrious life (Stabile 2012). These settlements were built around courtyard systems called "lotti" where paved and green spaces were dedicated to socialising and informal cultivation, a practice already widely accepted in the last century.

Over the years, processes of massive urbanisation have led to gentrification dynamics in neighbourhoods characterised by distinctive architectures. This has resulted in depopulation of the original settlement areas by inhabitants in favour of apartment owners, who, in the meantime, have been sold by the authorities to eligible individuals who rent or sell them at market prices.

This process has brought about a profound change in the social structure of the neighbourhood, transforming it through the emergence of more and more entities devoted to consumerism and entertainment, imbalanced compared to other more marginal areas that have become victims of abandonment and decay (Stabile 2012) (Fig. 6.1).

The Urban Gardens of Garbatella, one of the first experiences in the Roman territory, are the result of a long process of opposition to gentrification, cementification, and real estate speculation. They confirm the strong political connotation and social aggregation function of some forms of urban agriculture (Fig. 6.2).

In 2008, the residents of the neighbourhood decided to take action to defend an area that had never had a specific destination but was constantly at risk of being urbanised by private stakeholders or used as an illegal dump and then as a parking lot. Despite the substantial abandonment by institutions, the residents of the neighbourhood continued to carry out collective actions to protect the area, planting trees and trying to curb neglect.

The space began to come alive, especially when many nearby residents, including many retirees, approached the experience of reclaiming the area and began to participate in initiatives. The actions taken by this group of neighbourhood activists make explicit how these gardens constitute a strong symbol of active citizenship, based on a principle of decision-making autonomy in evaluating activities to shape the neighbourhood, starting with a clear rejection of real estate speculation (Del Monte and Sachsé 2017).

The initiative of the Garbatella urban gardens is part of a wider set of actions carried out since 2009 by the Legambiente circle of Garbatella, which, managed to involve a complex of Committees to address the task of formally managing the entire green area, according to an agreement proposed to the Municipality of Rome.

Fig. 6.1 Entrance to the Urban Gardens with the Lazio Region Building in the background. (Picture by Margherita Meta)

The realisation of the gardens is a fortunate result of a call for proposals from the Province of Rome for the regeneration of the area in 2010 and the need to continue to supervise the green area with initiatives of use and awareness; the Circle decided to propose to the Municipality of Rome a direct assignment contract for the management of the area, renewable annually (at no cost to the administration), also offering the commitment to create the park in the following years, according to the original project, reworked to include the gardens (Orti Urbani Garbatella 2024).

Initially, the activity of the urban gardeners was seen in a conflictual way by some residents of the neighbourhood, but despite the potentially emerging tensions it is possible to define a common profile regarding their activity within the gardens: almost all gardening members are involved in collateral activities of other social entities animating the area, demonstrating awareness of the importance of social cohesion. To participate in the gardens, it is necessary to apply to the Management Committee for the allocation of a plot of land. Once members of the garden, it is necessary to adhere to regulations that define commitments, rights, and the duration of the assignment, giving particular importance to the environment and to the value of social inclusion, expressed through the participation of associations working

Fig. 6.2 View of the vegetable gardens from the south side of the entrance. (Picture by Margherita Meta)

with migrants, disabled people, and former prisoners (Orti Urbani Garbatella 2024) (Fig. 6.3).

Significant centrality is attributed to the circular process of skills growth in approaching the territory. This process is brought to the forefront by continuous debate and discussion among garden members regarding the seeds to plant and the techniques to use, according to "a model of exchange that does not imply reciprocity but collective solidarity" (Del Monte and Sachsé 2017). Furthermore, the Garbatella urban gardens are not characterised as "municipal" gardens but rather as "community" gardens, not aimed at a single category of users as per common municipal regulations (usually the recipients are retired elderly), but at a very heterogeneous mix stimulating relationships: ordinary citizens, young unemployed or with precarious and occasional jobs, laid-off workers, retirees, former prisoners, in a stimulating intergenerational relationship, which fosters management based on sharing and the recovery of genuine relationships, experiencing a "different way of living where the collective good becomes a personal pleasure in itself" (Orti Urbani Garbatella 2024).

The experience of the Garbatella gardens combines social cohesion practices with the environmental sustainability to counter climate change. Inside the gardens,

Fig. 6.3 View of the honey hives, managed by the garden members. (Picture by Margherita Meta)

a rainwater collection system, a small photovoltaic system, and a composting system for the recycling of plant waste has already been installed and the use of pesticides and chemical fertilisers is prohibited.

This experience, fitting into a wider context of activism and claims for the re-appropriation of public space, fighting against real estate speculation and gentrification, stands out as extraordinarily unique for the role it plays as a social cohesive within the community (Fig. 6.4).

6.5 Conclusion

The practices of social cultivation are increasingly becoming a central theme in the context of urban governance agendas, particularly in relation to strategies for urban regeneration. The post-industrial outskirts of cities have served as a testing ground for the initial implementation of such practices (Gould and Tammy 2016). In social farming practices, the transformation of green spaces follows design criteria determined by participatory choices, reflecting a vision in which "the green turn in urban policy, planning, and design is rapidly emerging as a global orthodoxy guiding the way cities are reinvented and inhabited" (Newman 2015).

Similarly, the revitalization processes of the areas examined in this contribution have not only marked the re-appropriation and redevelopment of the green areas in question but also the emergence of new services, forms of citizen engagement in interaction with public spaces, and practices of urban renewal advocated by

Fig. 6.4 View of the vegetable gardens from the internal side. In the background, a public housing complex built in the 1920s. (Picture by Margherita Meta)

mechanisms of participatory governance and social inclusion. These practices highlight the need for planning to acknowledge social contradictions, in a manner detached from the established mode of local decision-makers, closely tied to the availability of economic resources and implementation timelines.

The lack of established models of participatory planning processes that centrally emphasise the importance of diverse visions should become the starting point for recognizing the value and potential of such experiences in relation to life in urban communities and a different relationship with food and health, with nature and the surrounding environment, with the future that we contribute to building with our way of living and governing.

These experiences can be implemented with a varied and progressive degree of richness and innovation to create social, environmental, and economic value and to build increasingly sustainable communities through the lifestyles and practices of its inhabitants and the policies of its administrators.

The investment required to implement urban horticulture plans seems capable of ensuring a significant return in terms of added value to the involved community. However, it is evident that this entails the challenge and necessity of adopting a visionary and innovative approach in managing such experiences. In return, this guarantees an extremely positive value in terms of impact on people's lives through a path destined to develop, with individuals and communities as co-protagonists.

Through this approach, collective experiments arise on urban agriculture realities that give space to reflections on the need to subvert the economic system based on the capitalization of resources and privatisation, giving rise to experiences that stem from an approach based on "an ethics not readable according to a scale of merely utilitarian values, as it is not directly aimed at producing market value" (Campiglio 2012).

These activism practices are generating new collective ways of inhabiting the urban context, responding to the crises of public space, and through bottom-up participation models, are initiating paths that do not conceive food and natural resources as simple objects with economically quantifiable value, but as common and collective heritage (Del Monte and Sachsé 2017).

Urban garden spaces also serve an important educational function for communities, incorporating various disciplines related to ecology, sociology, architecture, and geography, and serving as an educational laboratory that positively influences lifestyles through the social education of local communities. Therefore, it is increasingly necessary to implement local policies to integrate them into urban regeneration strategies to enhance the goal of reconnecting communities through good citizenship practices.

Thinking about the future of cities, it is necessary to reflect on the theme of urban and local agriculture not as a temporary remedy against emergencies but as a way to consider, on one hand, a more sustainable approach to designing and living in cities and, on the other hand, more complex food systems that better consider the specificities of places and the diverse needs expressed by current lifestyles (Sommariva 2012), as well as the integration of agriculture into urban and fringe spaces as an operational landscape for environmental sustainability and habitation.

Introducing standards for social agricultural practices would make a significant contribution to focusing on structural social issues, such as the marginalisation of peripheral areas, population ageing, and unemployment. Inhabitants of peri-urban agricultural landscapes are drawn to rural values while not wanting to give up the services and comforts offered by the city, thus proposing an innovative way to conceive the landscape in the social fabric of the widespread reality.

This type of perspective necessitates the search for new forms of governance and planning through a multidisciplinary approach that actively involves the public, institutions, and local stakeholders to implement participatory urban programs. In other words, a project that is built on the idea of improving the relationship between places and the communities residing in them, attempting to interpret the sense of the territory by reconfiguring it and attributing a central role to the potential for creating social bonds. Urban and peri-urban agriculture, as described above, implies a deeper rethink that goes beyond the concept of exclusively environmental protection in favour of policies that not only passively protect landscapes but also ensure the sustainability of their uses through appropriate management, intervening where there is degradation, critical issues, or profound ecological alteration (Sommariva 2012).

Participating in the deliberation of such practices is the most effective recipe for engaging those who are already active locally while working to integrate new social relationships into the process. Involving communities is the fundamental objective of learning, through collective actions that implement good practices, to create new horizons of active citizenship to combat marginalisation and build social cohesion.

In practical terms, in the implementation of spaces for social cultivation, this intent is achieved by closely correlating the relationship between agriculture and the city. In fact, despite the historically interdependent relationship they have always had, in the last century, drastic changes have occurred regarding factors such as transportation, production methods, and the rapid growth of the population. The increasing awareness of the unsustainable cost of deforestation, the realisation that the supply of fresh food can help combat food waste by composting waste and fertilising the soil, and the understanding that urban agriculture can create job opportunities, increase the possibility of community building, reduce outflow, and facilitate adaptation to climate change have led to a growing trend in the past 25 years for social cultivation practices to become a solid integral part of the urban environment.

Empty lots for cultivation are scarce in most large cities, but some innovative solutions, such as green roofs, can help create new opportunities for agriculture, allowing the establishment of cultivation spaces even in the absence of green spaces in urban areas, without compromising the creation of social bonds through good citizenship practices (Davids 2023).

References

Campiglio E (2012) L'economia buona. Milano-Torino, Bruno Mondadori

Contratto di comodato d'uso, art. 1803 Codice civile. https://gazzettaufficiale.it. Accessed 21 Feb 2024

Dansero E et al (2020) L'emergere del sistema cibo e delle sue fragilità nell'emergenza, Working papers. Online issue of Urban@it. https://doi.org/10.6092/unibo%2Famsacta%2F6790. Accessed 21 Feb 2024

Davids R (2023) Agripolis: living the high life, EDULEARN23 Proceedings, 1860–1867

Del Monte B, Sachsé V (2017) Coltivare la città Gli orti urbani condivisi come pratica di riappropriazione dello spazio pubblico nel contesto romano, in Antropologia, online issue of Cultural Anthropology, 4(3), Ledizioni, Milan

Gli Orti Urbani Garbatella. Metodo di intervento e caratteristiche distintive, http://www.legambientegarbatella.org/gli-orti-urbani-garbatella/ Accessed 21 Feb 2024

Gould K, Tammy L (2016) Green gentrification. Routledge, London

Harvey D (1984) On the history and present condition of geography: an historical materialist manifesto. Prof Geo 1(36) Taylor&Francis Online Library. https://www.tandfonline.com/doi/abs/10.1111/j.0033-0124.1984.00001.x

Milanesi F (2023) Orti Generali di Torino, un modello di rigenerazione urbana vince il Premio Nazionale del Paesaggio. https://touringclub.it. Accessed 21 Feb 2024

Newman A (2015) Landscape of discontent: Urban sustainability in immigrant Paris. University of Minnesota Press, Minneapolis

Orti Generali di Torino: il progetto. https://www.ortigenerali.it/agricoltura-urbana-nel-parco/. Accessed 21 Feb 2024

Paci A (2010) Prossimità e partecipazione. Lo spazio della sussidiarietà nella governance dei beni comuni. In Donati A. (a cura) Sussidiarietà e concorrenza. Una nuova prospettiva per la gestione dei beni comuni. Il Mulino, Bologna

Pascale A (2015) Gli Orti sociali in Italia. https://www.alfonsopascale.it/index.php/gli-orti-sociali-in-italia/. Accessed 21 Feb 2024

Petrulli M (2022) Gli orti sociali: finalità e disciplina regolamentare, in Outdoor: Living, Design, Technology 2, Maggioli, Rimini

Pettinati T., Toldo A., 2018, Il cibo tra azione locale e sistemi globali. Spunti per una geografia dello sviluppo, Franco Angeli, Milan

Pierri A, Torquati B (2016) Forme contrattuali e responsabilità nella gestione degli orti urbani, in Agriregionieuropa anno 12 n°44, Mar 2016. https://agriregionieuropa.univpm.it/. Accessed 21 Feb 2024

Reich RB (1991) The Work of Nations. Penguin Random House, New York

Ricci P (2007) Introduzione all'economia aziendale. Giuffrè, Milano

Sommariva E (2012) Agricoltura urbana: strategie per la città dopo la crisi, Atti XV conferenza Nazionale SIU. Planum J Urban 2(25) https://media.planum.bedita.net/24/82/12-XV%20Conferenza%20SIU%20by%20Planum%20n.25-2012%20Sommariva%20Atelier%206.pdf. Accessed 21 Feb 2024

Stabile FR (2012) La Garbatella a Roma. Architettura e regionalismo. Editrice librerie Dedalo, Roma

Vasile M, Grasseni C (2022) Visions of the urban green interrogating urban renewal in Turin's periphery, Anthrovision Vaneasa Online Journal, VANEASA—Visual Anthropology Network of European Association of Social Anthropologists. https://doi.org/10.4000/anthrovision.6812

Chapter 7
Urban Regeneration with Nature-Based Solutions: Human and Nature Perspectives

Katia Perini and Enrica Roccotiello

Abstract Urban ecosystems are complex, fragmented, densely populated areas, characterized by high level of anthropic pressure, lack of green spaces, habitat fragmentation and local biodiversity loss. In addition, most urban areas face important issues, among which the heat-island effect, poor air quality, and, overall, the deterioration of water resources and ecosystems. Sustainable urban regeneration represents a unique opportunity for city re-development with the aim to reinvent and rethink spaces, specifically underused, distressed, or decaying buildings or areas. Nature-based Solutions can play a significant role in such regeneration process, being able to provide various ecosystem services for the improvement of urban ecological and environmental quality. The present chapter describes the example of the Genoa (Italy) EU H2020 funded UNaLab pilot project and related monitoring activities, implemented in the field of human wellbeing and support to local biodiversity.

Keywords Biodiversity · Ecosystem services · Ecosystem disservices · Invasive alien species · Monitoring · Pilot project · Social benefits · Sustainability · Urban areas · Wellbeing

7.1 Introduction

Urban ecosystems are complex, fragmented, densely populated areas, characterized by high level of anthropic pressure, lack of green spaces, habitat fragmentation and local biodiversity loss (Grimm et al. 2000; Wu 2008). In addition, most urban areas

K. Perini (✉)
Department of Architecture and Design (DAD), University of Genoa, Genoa, Italy
e-mail: katia.perini@unige.it

E. Roccotiello
Department of Earth, Environment and Life Sciences, University of Genoa, Genoa, Italy
e-mail: enrica.roccotiello@unige.it

A. Battisti et al. (eds.), *Urban-Rural Dialogue in Green City Design*, Designing Environments, https://doi.org/10.1007/978-3-031-88318-7_7

face important issues, among which the poor air quality, heat-island effect and, overall, the deterioration of water resources and ecosystems (Di Giulio et al. 2009). Sustainable urban regeneration represents a unique opportunity for city re-development with the aim to reinvent and rethink spaces, specifically underused, distressed, or decaying buildings or areas (UN 2021).

Resilient and sustainable cities can be achieved through the restoration of ecological processes and functions, the increase of biodiversity in terms of quantity and quality (native species) and the equilibrium of the biological community (Capotorti et al. 2020; Pinho et al. 2021). The introduction of natural elements in urban environments allows the establishment of multitrophic interactions, that means interactions that link several trophic levels, including plants, herbivores, and predators, parasitoids, or pathogens (Brodie et al. 2018).

Specifically, biodiversity can be defined as the variability of living beings of all origins including, among others, aquatic ecosystems, and the ecological complexes they are a part of; this includes diversity within species and the diversity of ecosystems (UN 1992 Convention on Biological Diversity). In addition, several key international regulations were created to provide the knowledge and rules to manage, support, promote, restore biodiversity of natural and anthropic habitats like CE 92/43/EEC on the conservation of natural habitats and of wild fauna and flora (European Union 1992), World Summit on Sustainable Development (UN 2002), the European Green Deal (European Commission 2019), the EU Biodiversity Strategy for 2030 (European Commission 2020) and the nature restoration law (COM/2022/304) (IPBES 2022).

Since the two main causes of extinction are related to habitat loss and occurrence of invasive alien species (IPBES 2023), a key aspect at the urban level is the creation of habitat through Nature-based Solutions (NbS, i.e., actions to protect, sustainably manage, and restore natural or modified ecosystems providing benefits to human well-being and biodiversity (IUCN 2016) and the use of native plants to protect, restore and enhance the local biodiversity. To this aim an adequate planning should be consider the use of feeding plants for pollinators birds, and little mammals in NbS.

NbS can play a significant role in such regeneration process, being able to provide various ecosystem services (ES), i.e., the benefits people obtain from ecosystems (Millennium Ecosystem Assessment (Program) 2005). ES (Fig. 7.1) were divided into four main categories: provisioning, regulating, cultural and support services for a human-centred perspective (Maes et al. 2016). All the ES—like the provision of food, clean air, water, and resources as wood or medicine—can only be provided by healthy ecosystems and play a crucial role in supporting local livelihoods. NbS should be smartly designed avoiding the use or the occurrence of ecosystem disservices i.e., the negative effects of nature on human wellbeing (e.g., occurrence of invasive alien species, damage of buildings and infrastructures, and allergies) (von Döhren and Haase 2019; Tian et al. 2020).

Citizens show a keen interest in urban parks, viewing them as areas where they can connect with nature (Andersson et al. 2014). While utilizing an urban park, individuals may directly experience and appreciate certain ecosystem services, such as the cooling effect provided by plants (Mosca et al. 2021). However, they may not

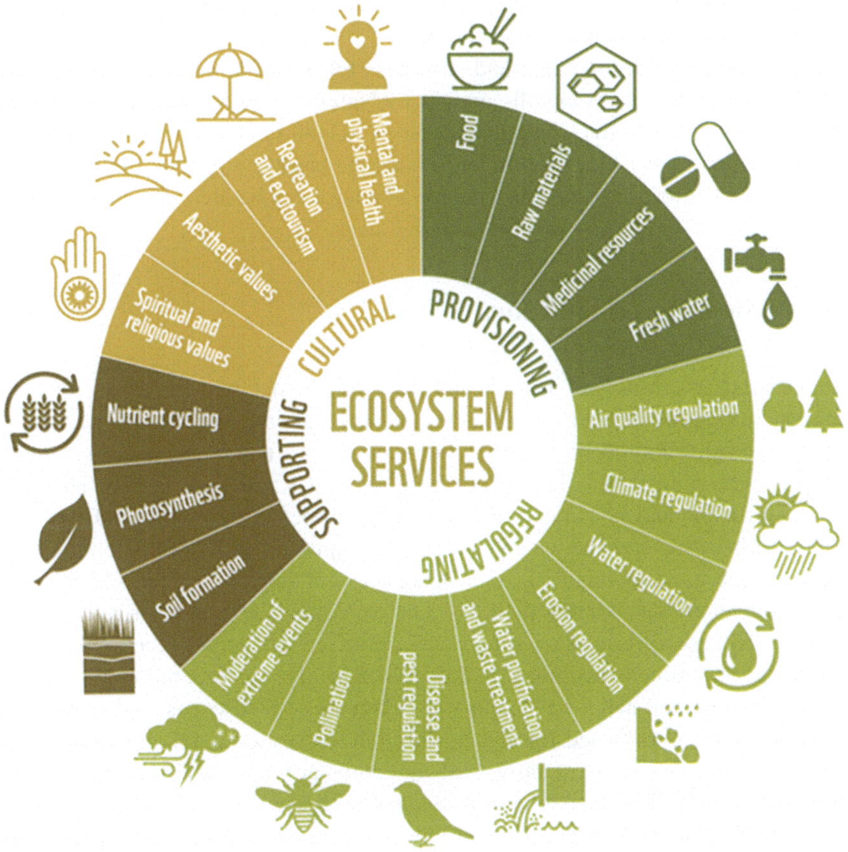

Fig. 7.1 Ecosystem services provided by nature to humans. WWF, 2016 (Adapted from Millennium Ecosystem Assessment 2005)

be immediately aware of other services, such as carbon sequestration and storage by park trees (Buchel and Frantzeskaki 2015).

While the benefits of urban ecosystems may not always be explicitly evident to city dwellers, it is essential to identify which ecosystem services are recognized by them. This understanding helps evaluate the perceived value and quality of existing urban green spaces. The belief is strong that fostering more sustainable cities involves not only enhancing the abiotic and biotic aspects of urban life but also addressing the social aspects, including people's satisfaction, experiences, and perceptions of the quality of their daily surroundings (Chiesura 2004).

In the past years, many projects were funded (e.g., through the H2020 programme) with the aim of reclaiming, redeveloping, and co-creating with local population abandoned urban spaces. Among them, the UNaLab project can be mentioned. The UNaLab project (UNaLab 2017), funded by the European Union's Horizon 2020 research and innovation programme aims at contributing to the "development of smarter, more inclusive, more resilient and increasingly sustainable cities through the implementation of nature-based solutions".

The chapter outlines strength and weakness related to urban areas in relations with the potential services and disservices by Nature based Solutions, considering the case study of one of the UNaLab pilot projects, built in Gavoglio former military barrack (Genoa, Italy) with the aim to evaluate the biodiversity and human wellbeing in the area before and after the creation of a large public area with NbS.

7.2 Nature-Based Solutions

Nature-based Solutions "are inspired and supported by nature", "are cost-effective, simultaneously provide environmental, social, and economic benefits and help build resilience. Such solutions bring more, and more diverse, nature and natural features and processes into cities, landscapes and seascapes, through locally adapted, resource-efficient and systemic interventions." (European Commission 2023).

NbS address societal challenges by drawing inspiration from and receiving support from nature, answering to the increasing needs of tangible solutions to increase urban biodiversity (Weisser et al. 2023). These solutions are characterized by their cost-effectiveness, offering concurrent environmental, social, and economic benefits while contributing to resilience-building. NbS involve the integration of a greater variety of nature and natural elements into urban areas, landscapes, and seascapes. This integration is achieved through locally tailored, resource-efficient, and systemic interventions. Consequently, NbS not only promote biodiversity but also enhance human well-being (Cohen-Shacham et al. 2016) (Fig. 7.2).

NbS provide a wide range of ecosystem services. According to (Babí Almenar et al. 2021), these solutions can be divided into: built structures, such as green roof or green wall, land media, as tree groups, meadows, orchard, and water media, as draining pavement, rain garden or infiltration basin. The ecosystem services provided by built structure mainly regard food provision (green roof especially), local climate and air quality regulations (Coma et al. 2018; Rowe 2018), carbon sequestration and storage, runoff mitigation (green roof) (Palla and Gnecco 2018). Green roof (Fig. 7.3) and green wall (Fig. 7.4) also provide habitat for species (Köhler and Ksiazek-Mikenas 2018; Pianta et al. unpublished) and improve recreation and mental and physical heath (Kotzen 2018). Tree groups and other land media NbS provide local climate and air quality regulations (Akbari et al. 2001; Lazzari et al. 2018), carbon sequestration and storage (Field et al. 2020), erosion prevention and maintenance of soil fertility (Gonzalez-Ollauri 2022), provide habitat for species (Atkins 2018), and improve recreation and mental and physical heath (Kim and Miller 2019). Finally, water media NbS, managed by e.g., raingarden (Fig. 7.5), play a key role in the moderation of extreme events and runoff control/mitigation (Ballard et al. 2007; Woods Ballard et al. 2015).

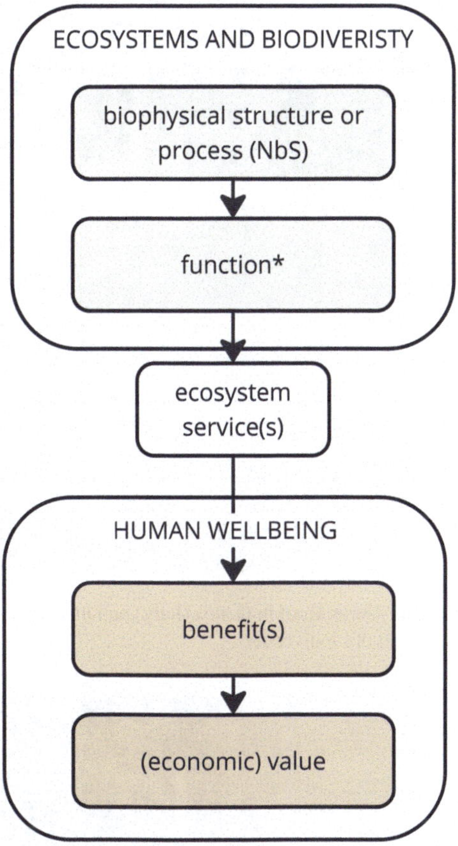

Fig. 7.2 Framework linking ecosystem to human wellbeing (de Groot et al. 2010), modified. *function: a subset of biophysical structure or process providing the service

7.3 Quantifying the Performances of Nature-Based Solutions: The Case of UNaLaB

The UNALAB-URBAN NATURE LABS (UNaLab) project, funded by the European Program "HORIZON 2020" call. Smart and sustainable Cities—SCC—02-2016-2017: Demonstrating innovative Nature-based Solutions in Cities (topic SCC-02-2016-2017 type of action IA), aims at developing smarter, more inclusive, more resilient, and increasingly sustainable cities through the implementation of NbS.

One of the three front runner cities of the project is Genoa (Italy). Genoa, with approximately 600.000 residents, features a slender coastal area bordered by hills and steep mountains in the hinterland. The city encounters various environmental hurdles linked to frequent flooding which has resulted in significant destruction in the past, primarily due to intense rainfall on a highly urbanised landscape, including challenges in water management, heat stress, and issues of water and air pollution.

Fig. 7.3 Green Roof in Genoa (Italy) on DICCA, University of Genoa, with low management and high level of biodiversity

Fig. 7.4 Living wall on Giambellino's district, Milan (Italy) for particulate matter capture (H2020, Clever cities project)

Fig. 7.5 Rain Garden in Campomorone (Genoa, Italy), pilot project of Proterina3Evolution (Interreg IT-FR), photo Gianluca Porcile

Fig. 7.6 Gavoglio Urban Park with 12 NbS in the city of Genoa (Italy, 2023), northern area

The UNaLaB project partially funded the construction of a new multifunctional public park on the site with 12 different NbS in the former Gavoglio barracks, located in the Lagaccio district, a densely populated residential area (UNaLab 2017), in an intensely artificialized urban environment (Figs. 7.6 and 7.7).

Fig. 7.7 Gavolgio Urban Park with 12 NbS in the city of Genoa (Italy, 2023), southern area activities are implemented to evaluate the performances in terms

UNaLaB Genoa pilot project monitoringof ecosystem services provision. An interdisciplinary group is monitoring the performances of NbS in improving ecological, environmental, and economic condition of a district. Monitoring and data collection is related to the key performance indicators:

- Mitigation of the Urban Heat Island effect
- Capture of PM_{10}, $PM_{2.5}$, NO_2 and O_3 in ambient air
- Citizen wellbeing associated with environmental conditions (microclimate, psychological wellbeing, social cohesion)
- Active involvement of citizens
- Real estate value increase of buildings
- Biodiversity (plants)
- Pollinator insects
- Evapotranspiration rate
- Carbon sequestration
- Stormwater retention
- Water used for irrigation.

In this chapter, we focus on the key performance indicators listed in Table 7.1.

Table 7.1 UNaLaB Genoa pilot project monitoring activities developed by the University of Genoa in the field of Architecture/psychology, Environmental botany, Key Performance Indicators, and related measurement unit in relations with ecosystem services

Key Performance Indicators	Measure unit	Ecosystem services
Citizen wellbeing associated with environmental conditions (microclimate, psychological wellbeing, social cohesion)	N° of people, % satisfaction	**Cultural services** Recreation and mental and physical health Aesthetic appreciation and inspiration for culture, art, and design Social cohesion
Biodiversity	Shannon's index (H′), F%, NP/P	**Habitat or supporting services** Species richness and diversity (plants and birds)
Pollinator insects	Shannon-wiener index (H′)	**Habitat or supporting services** Pollination
Evapotranspiration	ET_0 (mm/day)	**Regulating services** Local climate and air quality regulation
Carbon sequestration	% increase; t/ha/year	**Regulating services** Carbon sequestration and storage

7.4 Citizen Wellbeing

Analysing the public's perception and enhancing well-being associated with the utilization of NbS in public areas involves examining how individuals perceive these solutions at both the individual and community levels. This assessment is crucial for gauging the positive impact on human well-being resulting from the ability to enjoy green spaces within urban environments (Chiesura 2004). The monitoring process related to the KPI Citizen wellbeing associated with environmental conditions aims to evaluate, both qualitatively and quantitatively, citizens' perceptions of ecosystem services provided by various NbS. It considers the levels of well-being and satisfaction expressed concerning the characteristics of the urban context. To conduct this assessment, data were gathered through a questionnaire, compiled by approximately 600 citizens living in the area. In this chapter we focus on the public perception of ecosystem services provided by the NbS. More specifically, we consider the relationship man-nature within the city and perceived performances of a selected number of NbS, among the ones used in the Genoa pilot project:

- Raingarden
- Infiltration basin
- Tree groups
- Drought-resilient orchard and meadows
- Green wall

- Natural engineering for slope securing.

The results of the monitoring activity show that the function of NbS is generally recognized as positive, specifically for tree-lined areas, rain gardens and infiltration basins. In the case of green walls, the most recognized effect is reduction of urban heat island (ecosystem service microclimate regulation). Also, in the case of tree groups this is the most recognized effect, along with increase of biodiversity, and rainwater management. Rain gardens are considered more for biodiversity increase than for rainwater management. This latter is correctly considered by most of the citizens as main benefit of infiltration basin.

Finally, different issues have been raised by local population: from one side having more access to green spaces for leisure and higher demand for more liveable cities, on the other, a lack of recognition of the potential environmental and ecological benefits of the park, respect to the potential ones (e.g., biodiversity improvement, CO_2 capture, etc.), was noticed.

7.5 Biodiversity

Alteration of the biodiversity in terms of species occurrence, richness (and related network) and ecological functions in the urban context is frequent, and result in a disruption and decrease of different living species (Hillebrand et al. 2018). In addition, the occurrence of exotic species that are or can become invasive (because of the absence of predators, the availability of food and the increasing temperatures) could affect the performances and effectiveness of NbS. The pre-opera monitoring on the whole area, that was subsequently reclaimed, was done to evaluate the state-of-the-art and the risk of occurrence of invasive alien species.

The monitoring after park construction was focused on the NbS realized after the reclamation process and specifically: raingarden, infiltration pond, bioswale, tree groups and green areas, drought-resilient orchard and meadows, slope afforestation, green wall, and natural engineering for slope securing. For all these NbS, the plant diversity was evaluated with the Shannon-Weiner Index (H′), that is one of the most widely used for this purpose (Shannon 1948).

The monitoring preceding and during park construction showed a total of 91 *taxa*, 73 of which are autochthonous (indigenous) and 18 allochthonous (exotic). Allochthonous species represent all those species present in an area that does not correspond to their regular range of distribution, but they have been dislocated there by volunteer or inadvertent human actions. Among the 18 alien species recorded, 12 are invasive, and 6 non-invasive, where invasives optimally adapt to specific habitat out of their natural range, causing damage to the whole ecosystem. Considering the high ratio between invasive and non-invasive species recorded we can consider these results coherent with the conditions of the area, typical of a construction site with excavated and incoherent terrain, where the previous balances were modified due to the continuous human disturbance. The occurrence of exotic, invasive or potentially invasive species can eventually affect the effectiveness of the NbS to be realized, thus it is important to list them to plan eradication or mitigation actions.

The first monitoring, immediately after the park opening, showed that plant biodiversity, considering a pre-existing plant coverage reduced to a minimum and an area recently planted and seeded, consists of 52 *taxa*. The decrease in the number of plant *taxa* is related to the eradication of invasive species and establishment of new plant communities. The choice of numerous native herbaceous and shrub species, typical of the Mediterranean environment, and the replacement of the forest species envisaged by the project with some forest trees more coherent in terms of climate and representing pre-existing plant populations, will result in more resilient plant communities, able to better resist to environmental stresses. As regards the plant biodiversity indices, they are NbS-dependent because of the species choice and related coverage, and relatively high, but details on invasive species occurrence should complement such findings in the future. A subsequent monitoring with respect to time zero (park opening) is recommended to evaluate a possible increase, remembering that natural vegetation dynamics will establish in decades.

7.6 Summary and Future Perspectives

The monitoring of NbS is a key aspect that allow evaluating the performances of such solutions in relation to a specific context. In the portrayed case study, we monitored NbS performances based on small-scale indicators that can be used to evaluate ecosystem services provision in a degraded urban area (a former military barracks). This approach was implemented by Genoa, Tampere and Eindhoven, i.e. the front-runner cities of the UNaLaB project, and also by the seven following cities, to help creating the knowledge base for developing a standardized set of indicators for the implementation of urban NbS that can be used at EU level, summarised in the "Nature-Based Solutions Implementation Handbook" (Dubovik et al. 2022).

What emerges from the monitoring activities is that the construction of the new park on the abandoned Gavoglio Barracks has overall a positive effect on the surrounding area. The citizens participating in the monitoring activity show an adequate knowledge of the ecosystem services provided by the single NbS analysed.

Plant diversity recorded before and after NbS realization highlighted the importance of a constant monitoring since natural dynamics require time for a more inclusive evaluation of such structure in terms of quantity and quality of the results through time, thereby providing data to manage site colonization by invasive species in the future.

It is well known that urban biodiversity positively impacts citizens' wellbeing by enhancing mental health, providing recreational spaces, improving air quality, and fostering social connections. Urban parks, like the one presented in this chapter, provide diverse ecosystems in cities, and offer residents opportunities for relaxation, exercise, and interaction with nature, leading to healthier, happier, and more resilient urban communities.

According to these results, it is possible to assume that the positive effects of the UNaLab Genoa pilot project are well understood. Further monitoring activities are under implementation to assess the perception of users in terms of wellbeing and biodiversity improvement in different areas of the UNaLab pilot project built in Genoa (construction was completed in August 2022), over time. Indeed, the perception of citizens will change in the coming months and years as the vegetation grows and the park becomes more established. Further monitoring activities are even more important when speaking about biodiversity.

Since vegetation is naturally dynamics and communities tend to evolve and change over time, monitoring the biodiversity of NbS in the park must be conducted consistently over time to understand if and how the area's biodiversity is changing, identifying the main drivers of change. This information will allow for differentiated and sustainable management of the area, creating a stable network of living organisms, also benefiting the resident population.

In addition, care should be taken where the urban green areas, like parks, are surrounded by native vegetation or are located closed to urban-rural transition areas. Currently, there is still a lack of information on how plant communities respond to intermediate, persistent, and widespread disturbances attributable anthropogenic pressures. The latter can affect biotic communities (i.e., plants, animals, and humans) and their interactions, resulting in a reshaping of originally constructed green areas.

Acknowledgments This work was supported by the UnaLab project, funded by the European Union's Horizon 2020 research and innovation programme under Grant Agreement No. 730052, under topic SCC-022016-2017 "Smart Cities and Communities Nature based solutions".

References

Akbari H, Pomerantz M, Taha H (2001) Cool surfaces and shade trees to reduce energy use and improve air quality in urban areas. Sol Energy 70:295–310. https://doi.org/10.1016/S0038-092X(00)00089-X

Andersson E, Barthel S, Borgström S et al (2014) Reconnecting cities to the biosphere: stewardship of green infrastructure and urban ecosystem services. Ambio 43:445–453. https://doi.org/10.1007/s13280-014-0506-y

Atkins E (2018) Chapter 3.15: Green streets as habitat for biodiversity. In: Pérez G, Perini K (eds) Nature based strategies for urban and building sustainability. Butterworth-Heinemann (Elsevier), Oxford, pp 251–260

Babí Almenar J, Elliot T, Rugani B et al (2021) Nexus between nature-based solutions, ecosystem services and urban challenges. Land Use Policy 100:104898. https://doi.org/10.1016/j.landusepol.2020.104898

Ballard BW, Kellagher R, Martin P et al (2007) Site handbook for the construction of SUDS. CIRIA

Brodie JF, Redford KH, Doak DF (2018) Ecological function analysis: incorporating species roles into conservation. Trends Ecol Evol 33:840–850. https://doi.org/10.1016/j.tree.2018.08.013

Buchel S, Frantzeskaki N (2015) Citizens' voice: a case study about perceived ecosystem services by urban park users in Rotterdam, the Netherlands. Ecosyst Serv 12:169–177. https://doi.org/10.1016/j.ecoser.2014.11.014

Capotorti G, Bonacquisti S, Abis L et al (2020) More nature in the city. Plant Biosyst Int J Deal Aspect Plant Biol 154:1003–1006. https://doi.org/10.1080/11263504.2020.1837285
Chiesura A (2004) The role of urban parks for the sustainable city. Landsc Urban Plan 68:129–138. https://doi.org/10.1016/j.landurbplan.2003.08.003
Cohen-Shacham E, Walters G, Janzen C, Maginnis S (eds) (2016) Nature-based solutions to address global societal challenges. IUCN International Union for Conservation of Nature
Coma J, Pérez G, Cabeza LF (2018) Chapter 3.2: Green roofs to enhance the thermal performance of buildings and outdoor comfort. In: Pérez G, Perini K (eds) Nature based strategies for urban and building sustainability. Butterworth-Heinemann (Elsevier), Oxford, pp 109–117
de Groot RS, Alkemade R, Braat L et al (2010) Challenges in integrating the concept of ecosystem services and values in landscape planning, management and decision making. Ecol Complex 7:260–272. https://doi.org/10.1016/j.ecocom.2009.10.006
Di Giulio, M., Holderegger, R., & Tobias, S. (2009). Effects of habitat and landscape fragmentation on humans and biodiversity in densely populated landscapes. Journal of Environmental Management, 90(10), 2959–2968.
Dubovik M, Rinta-Hiiro V, Laikari A, et al (2022) Nature-based solutions implementation handbook
European Commission (2019) Communication from the commission to the European Parliament, the European Council, the Council, the European Economic and Social Committee and the Committee of the Regions. The European Green Deal
European Commission (2020) Communication from the Commission to the European Parliament, the Council, the European economic and social committee and the committee of the regions EU biodiversity strategy for 2030 bringing nature back into our lives
European Commission (2023) Nature-based solutions. https://research-and-innovation.ec.europa.eu/research-area/environment/nature-based-solutions_en. Accessed 6 Dec 2023
European Union (1992) COUNCIL DIRECTIVE 92/43/EEC on the conservation of natural habitats and of wild fauna and flora
Field RH, Buchanan GM, Hughes A et al (2020) The value of habitats of conservation importance to climate change mitigation in the UK. Biol Conserv 248:108619. https://doi.org/10.1016/j.biocon.2020.108619
Gonzalez-Ollauri A (2022) Sustainable use of nature-based solutions for slope protection and erosion control. Sustain For 14:1981. https://doi.org/10.3390/su14041981
Grimm NB, Grove JG, Pickett STA, Redman CL (2000) Integrated approaches to long-term studies of urban ecological systems: urban ecological systems present multiple challenges to ecologists—pervasive human impact and extreme heterogeneity of cities, and the need to integrate social and ecological approaches, concepts, and theory. Bioscience 50:571–584. https://doi.org/10.1641/0006-3568(2000)050[0571:IATLTO]2.0.CO;2
Hillebrand H, Blasius B, Borer ET et al (2018) Biodiversity change is uncoupled from species richness trends: consequences for conservation and monitoring. J Appl Ecol 55:169–184. https://doi.org/10.1111/1365-2664.12959
IPBES (2022) Global assessment report on biodiversity and ecosystem services of the intergovernmental science-policy platform on biodiversity and ecosystem services. Zenodo. https://zenodo.org/record/5657041
IPBES (2023) Thematic assessment report on invasive alien species and their control. IPBES secretariat
IUCN (2016) Nature-based solutions. In: IUCN. https://www.iucn.org/commissions/commission-ecosystem-management/our-work/nature-based-solutions. Accessed 20 Sep 2021
Kim G, Miller PA (2019) The impact of green infrastructure on human health and well-being: the example of the Huckleberry Trail and the Heritage Community Park and Natural Area in Blacksburg, Virginia. Sustain Cities Soc 48:101562. https://doi.org/10.1016/j.scs.2019.101562
Köhler M, Ksiazek-Mikenas K (2018) Chapter 3.14: Green roofs as habitats for biodiversity. In: Pérez G, Perini K (eds) Nature based strategies for urban and building sustainability. Butterworth-Heinemann (Elsevier), Oxford, pp 239–249

Kotzen B (2018) Chapter 4.2: Green roofs social and aesthetic aspects. In: Pérez G, Perini K (eds) Nature based strategies for urban and building sustainability. Butterworth-Heinemann (Elsevier), Oxford, pp 273–281

Lazzari S, Perini K, Roccotiello E (2018) Chapter 3.6: Green streets for pollutants reduction. In: Pérez G, Perini K (eds) Nature based strategies for urban and building sustainability. Butterworth-Heinemann (Elsevier), Oxford, pp 149–156

Maes J, Zulian G, Thijssen M, et al (2016) Mapping and assessment of ecosystems and their services—urban ecosystems 4th report

Millennium Ecosystem Assessment (Program) (2005) Ecosystems and human well-being: synthesis. Island Press, Washington, DC

Mosca F, Dotti Sani GM, Giachetta A, Perini K (2021) Nature-based solutions: thermal comfort improvement and psychological wellbeing, a case study in Genoa, Italy. Sustainability 13:11638. https://doi.org/10.3390/su132111638

Palla A, Gnecco I (2018) Chapter 3.11: Green roofs to improve water management. In: Pérez G, Perini K (eds) Nature based strategies for urban and building sustainability. Butterworth-Heinemann (Elsevier), Oxford, pp 203–213

Pinho P, Casanelles-Abella J, Luz AC et al (2021) Research agenda on biodiversity and ecosystem functions and services in European cities. Basic Appl Ecol 53:124–133. https://doi.org/10.1016/j.baae.2021.02.014

Rowe B (2018) Chapter 3.5: Green roofs for pollutants' reduction. In: Pérez G, Perini K (eds) Nature based strategies for urban and building sustainability. Butterworth-Heinemann (Elsevier), Oxford, pp 141–148

Shannon CE (1948) A mathematical theory of communication. Bell Syst Tech J 27:379–423

Tian Y, Wu H, Zhang G et al (2020) Perceptions of ecosystem services, disservices and willingness-to-pay for urban green space conservation. J Environ Manag 260:110140. https://doi.org/10.1016/j.jenvman.2020.110140

UN 1992 United Nations. (1992). Convention on Biological Diversity. New York

UN (2002) Report of the World Summit on Sustainable Development, Johannesburg, South Africa, 26 August-4 September 2002

UNaLab (2017) Urban nature lab—H2020. https://unalab.eu/en. Accessed 22 Sep 2021

UN 2021. United Nations Human Settlements Programme (UN-Habitat). (2021). Urban Regeneration as a Tool for Inclusive and Sustainable Recovery. pp 120, UN-habitat Publisher

von Döhren P, Haase D (2019) Risk assessment concerning urban ecosystem disservices: the example of street trees in Berlin, Germany. Ecosyst Serv 40:101031. https://doi.org/10.1016/j.ecoser.2019.101031

Weisser WW, Hensel M, Barath S et al (2023) Creating ecologically sound buildings by integrating ecology, architecture and computational design. People Nat 5:4–20. https://doi.org/10.1002/pan3.10411

Woods Ballard B, Wilson S, Illman S, et al (2015) The SuDS manual

Wu J (2008) Making the case for landscape ecology: an effective approach to urban sustainability. Landsc J 27:41–50. https://doi.org/10.3368/lj.27.1.41

Chapter 8
Regenerative Foodscapes as a Holistic and Integrated Approach for Ecological Transition in the Alpine Territories

Sara Favargiotti and Angelica Pianegonda

Abstract Approximately 60% of global ecosystem services are currently degraded or unsustainably managed. Addressing this, Trento's initiative under the EIT Food research projects focuses on revitalizing abandoned and underutilized open areas, creating productive and recreational landscapes that promote agroecology. These efforts aim to establish inclusive, sustainable, and circular habitats, mitigating climate change risks through natural assets, and enhancing socio-ecological and biodiversity values in peri-urban areas. The research delves into the multifunctionality of urban and peri-urban agriculture in Trento. It highlights its role in providing environmental, economic, and social benefits, such as creating barriers against urban expansion, improving water management, regulating microclimates, diversifying natural environments, and boosting local biodiversity. Furthermore, it serves as a means for local food production, providing employment opportunities and supporting ecosystem services.

Keywords Foodscapes · Agroecology · Regenerative design · Sustainable food systems · Ecosystem services

8.1 Introduction

As the United Nations projects the global population to reach 8.5 billion by 2030 and 9.7 billion by 2050 (UN 2017), an inevitable surge in food demand looms on the horizon. This escalation poses critical questions about our food production methods and their environmental and societal impacts. In this milieu, cities, which currently house over half of the world's population and are expected to accommodate more than 75% by 2050, emerge as pivotal players. The challenge lies in reducing the

S. Favargiotti (✉) · A. Pianegonda
Department of Civil, Environmental and Mechanical Engineering, University of Trento, Trento, Italy
e-mail: sara.favargiotti@unitn.it; angelica.pianegonda@unitn.it

A. Battisti et al. (eds.), *Urban-Rural Dialogue in Green City Design*, Designing Environments, https://doi.org/10.1007/978-3-031-88318-7_8

length and complexity of the supply chain and transitioning towards more sustainable food systems to diminish the human environmental footprint.

The past few decades have seen the food system evolve into a highly industrialized entity (Barkema et al. 1993; Steel 2013), characterized by extensive supply chains and practices detrimental to the environment, such as using pesticides, chemicals, and fertilizers. While this system has played a role in reducing global malnutrition, it has also perpetuated socioeconomic disparities and environmental degradation (Rosset 2008; FAO 2011). Predicated on standardization and economic efficiency (Gaitán-Cremaschi et al. 2019), the current food system's interconnected nature means that alterations in one segment can have far-reaching effects.

Recent global events, including the pandemic and the Ukrainian war, have drastically altered food demand and supply, prompting governments to impose import and export restrictions (Espitia et al. 2020) and highlighting the role of social innovation in adapting food provisioning systems (Cattivelli & Rusciano 2020). These market fluctuations have exposed the vulnerabilities and paradoxes of the existing system (Ahn and Norwood 2021; Fitzpatrick et al. 2020). Therefore, there exists a pressing need for a food system that is more resilient, equitable, and sustainable, transforming current crises into opportunities for systemic change.

Key to this transformation is bridging the gap between urban and rural planning and governance, raising public awareness about the importance of preserving natural ecosystems and ecosystem services. While urban food system dynamics are increasingly entering public discourse, the influence of cities in shaping agricultural areas and foodscapes remains underexplored (Fig. 8.1). Transforming the industrialized food system into a more sustainable model presents an opportunity for social innovation, reconnecting citizens with their land, fostering community belonging, spurring territorial development, and enhancing urban liveability. The food system

Fig. 8.1 City-countryside relationship in Brasov, Romania. Ph. Angelica Pianegonda, 2022

can thus become a catalyst for territorial planning (Favargiotti and Pianegonda 2021), driving innovation and engaging citizens more actively.

By involving citizens, we can promote environmentally conscious lifestyles, increase demand for healthy, fresh, and local products, and thereby influence food production methods. Such a shift can enhance urban food security and sovereignty, improve public health, and lead to more sustainable natural resource management. Ultimately, cities surrounded by natural areas could see an increase in ecosystem services, thereby elevating the quality of life for their inhabitants. Recognizing the value of sustainable food production is crucial in advocating for the transformation of the current industrial food system into one that is healthier and more respectful of both the environment and society. This paper explores how regenerative foodscapes could offer a holistic and strategic framework for participatory design aiming to transform cities and territories through the implementation and integration of sustainable food ecosystems into spatial planning and landscape design and management for the ecological transition.

8.2 Regenerative Foodscapes for Restoring, Regenerating, and Reinhabiting the Existing

To develop landscape projects coherent with local and harmless food production, it is important that beneficial functions deriving from respectful management of agricultural areas are collectively recognized and bottom-up support is given. A first benefit linked to urban and peri-urban agriculture is, therefore, related to its environmental and ecological functions (Fig. 8.2), among which are the creation of space barriers against urban expansion; water management; microclimatic

Fig. 8.2 Biodiversity in the OrtoAperto Community Garden in Trento, Italy. Ph. Angelica Pianegonda, 2024

regulation; diversification of the natural environment and improvement of local biodiversity; and management of agricultural landscape structures: windbreak hedges, wells, canals, paths, vegetated and aquatic surfaces (Donadieu et al. 2013). In this sense, urban and peri-urban agriculture assume the role of the green infrastructure of the public interest. A second benefit deriving from urban and peri-urban agriculture is linked to its economic and social function thanks to the provision of goods and employment opportunities (Zasada 2011; Branduini et al. 2016). Urban and peri-urban agriculture gives consumers the chance to find local and fresh food and directly verify the origin of the products and how they are produced (Opitz et al. 2016). Moreover, it can be both a main or secondary income, also linked to indirect economic impacts, for example, linked to the gastronomic tourism sector (Hjalager and Richards 2003).

Moreover, urban, and peri-urban agriculture could provide individual benefits in terms of mental and physical health (Fig. 8.3), given the symbolic and aesthetic values related to these natural areas (Brown and Jameton 2000). In recent years, citizens have shown the willingness to experience more contact with nature since, in an increasingly dematerialized economy and society, proximity to the city is no longer essential to have access to work, goods, and services (Giudici 2021). The pandemic seems to have reinforced these dynamics (Nathan and Overman 2020) and highlighted the importance of open-air and natural spaces close to cities.

Fig. 8.3 OrtoAperto Community Garden activities. Ph. Angelica Pianegonda, 2024

Urban and peri-urban agriculture are, therefore, the bearers and developers of multifunctionality of spaces (Lovell 2010), services and benefits (Zasada 2011; Donovan et al. 2011), which can go far beyond those described above. Despite that, agricultural areas, specifically those surrounding cities, are the most affected places by transformation processes as they are not considered infrastructures of public interest and the added values provided by foodscapes are not recognized (Donadieu et al. 2013). Agricultural areas can no longer be managed by considering only their potential commercial value. Therefore, the mechanism of land rent and the surplus value generated by land-use change must stop to ensure the permanence of voids and overcome the logic of competition between the uses of space (Donadieu et al. 2013). Nevertheless, foodscapes (MacKendrick 2014) have often been opportunities to trigger processes of urban regeneration and rehabilitation of often abandoned or underutilized areas while offering opportunities for the renewal of the dichotomous relationship between urban and rural landscapes, the generation of new sustainable productive practices and the creation of a sense of community. Foodscapes must be intended as spatial, cultural, and social entities that guarantee urban well-being and a regenerative enhancement of natural and cultural resources. For this reason, they must be consistently and precisely integrated in urban plans and landscape projects. To achieve this goal, it is necessary to overcome the city-countryside antagonism that has characterized the development of regulatory plans until now and move the discussion from the concept of "land" to "territory" (Donadieu et al. 2013), thus giving more value to the cultural and ecological aspects and recognizing farmers as custodians of the landscape. For example, public administrations could protect these areas by paying for the ecosystem services they provide or de-taxing sustainable agricultural production. Local food policies could support this change by taking the landscape project as an operational tool to activate the spatial transformation of both urban and rural areas, including a more sustainable food system. In this framework, the agroecological approach can play a key role in reconnecting cities and their surroundings, recovering territorial biodiversity, producing healthy food, and promoting social innovation (Bocchi 2020; Gliessman 2018; Snapp 2017).

The regenerative foodscapes approach copes with the emerging need to move from top-down governance approaches to more inclusive processes, participative actions, and shared decisions (Méndez et al. 2013) as seen in the rise of urban-rural initiatives for food provisioning during the COVID-19 pandemic (Cattivelli & Rusciano 2020). The investigated case studies represent bottom-up spaces of experimentation in the management of food and implementation of collective areas, meaning a shift from over-imposed choices to community-based and co-designed processes of space transformation (Chambers 1994). The participatory design approach has spread since the 1970s, aiming to democratize decision-making processes, and has been recognised as a helpful tool to evaluate greater territorial and societal transformation by directly involving citizens (Luck 2018). Participatory design processes aim to stimulate the community to inhabit their territories, increase the sense of belonging, and take care of public spaces (Keil and Kistmann 2016). In recent decades, citizens have claimed greater participation (Fung 2015) in social and ecological justice (White 2007) in response to the increasing privatization of cities with top-down management of spaces, housing, and services (White 2007). These needs have been reflected in spatial justice that seeks to promote participatory approaches, citizen engagement, social activism, and the mobilization of grassroots initiatives. Interest in the food system has spread, developing Alternative Food Networks, and promoting the proliferation of community gardens and public vegetable gardens as "urban commons" (Colding et al. 2013). In this framework, food can be seen as an element that helps to manage public areas through inclusive approaches. In most of the XIX and XX centuries' urban plans, food was not considered in the design of cities (Marot 2019). In this framework, agricultural lands were — and in many cases are still — considered as potential areas of urban expansions where their productive, socio-economic, and environmental role is not considered (Donadieu et al. 2013). Only recently has urban and peri-urban agriculture entered the planning debate (Lovell 2010), as it has proven to respond to the challenges posed by urbanization, e.g., concerning food sovereignty and security, and as a resource for biodiversity in urban and peri-urban areas.

Moreover, urban agriculture can be considered a complementary income source, a social and cultural element, and it can be understood as "the common denominator in which a local community can identify and express itself" (Favargiotti and Pianegonda 2021). Despite the many positive aspects offered by urban agriculture, citizens are still disconnected from food production and, in general, from nature, reflecting a lack of awareness, and therefore of interest, in the sustainable management of natural resources and the management of rural areas. In urban contexts, commons offer important insights in response to economic and social crises (de Angelis 2012; Harvey 2012) and represent political arenas in which collective actions and processes of social emancipation can take place (de Angelis 2012; Harvey 2012). Given these premises, the concept of the commons has expanded to quite unexplored fields, as in the case of food as a common (Sumner 2011), to pursue social, political, and environmental transformations (Pelenc et al. 2019; Moreira and Morell 2020). Food has been increasingly recognised as a helpful tool to evaluate territorial and societal transformation by directly involving citizens, stimulating the community to have a deeper knowledge of their territories, and making citizens aware of the urgent need to safeguard the entire natural ecosystem. If the role of food as a tool to engage citizens in shaping spaces and dynamics in the urban and peri-urban context is well recognized, less discussed are the impacts and effects that it has on the territorial scale. In this contribution, food and foodscapes are investigated as common goods at the edge of urban and rural commons (Ferretti and Favargiotti 2022; Favargiotti et al. 2023) and as tools to reconnect urban and rural landscapes. The case study will be presented and discussed in the following paragraphs.

Urban agriculture in Europe plays a pivotal role in enhancing ecosystem services, contributing significantly to sustainable urban development. In Bologna, Italy, a study highlighted the social acceptance and perceived benefits of urban agriculture, revealing a strong preference among citizens for plant-based agricultural products. This preference underscores the cultural and social ecosystem services urban agriculture provides, fostering community engagement and promoting green spaces within urban settings (Sanyé-Mengual et al. 2018). Additionally, research in the Czech Republic has shed light on the adverse effects of urban sprawl on agricultural lands, emphasizing the loss of valuable ecosystem services. This study underscores the need for sustainable urban planning to preserve the ecological and economic benefits that these lands offer (Vrublova 2020).

Moreover, the geographical perspective of urban agriculture in Europe reveals its multifaceted role in urban and rural transformations, urban regeneration, and agricultural restructuring. It contributes to multifunctionality, providing ecosystem services, resolving land-use conflicts, and enhancing social responsibility (The geography of urban agriculture: New trends and challenges 2017). Furthermore, a bibliometric review aligning multifunctional agriculture with the Sustainable Development Goals (SDGs) indicates a growing research interest in Europe, focusing on ecosystem services, biodiversity, governance, and urban agriculture. This research trajectory highlights the integral role of urban agriculture in achieving sustainability objectives (Esquivel-Marín et al. 2023). Collectively, these studies demonstrate the critical importance of urban agriculture in European cities, not only for its direct provisioning services but also for its broader contributions to ecological sustainability and social well-being.

8.3 The City of Trento: A Case Study to Understand the Impact of the Food System on the Alpine Landscape

Trento is a municipality situated in the northern part of Italy and represents one of the urban centers in the Alpine region. The city, nestled in the mountains, serves as the capital of the Trentino province (Fig. 8.4). The region's topography is predominantly mountainous, this feature profoundly influences local spatial dynamics. Agriculturally, Trento presents a diverse and dynamic sector. The region's agricultural landscape is a mosaic of traditional and modern practices, deeply influenced by its alpine geography. Approximately 34% of Trento's territory is dedicated to agriculture, a sector characterized by a mix of small-scale farms and larger agricultural enterprises. Notably, the region hosts around 7800 farms, with nearly 3.000 operating on areas smaller than 0.5 ha (Andreola et al. 2021). This reflects a trend

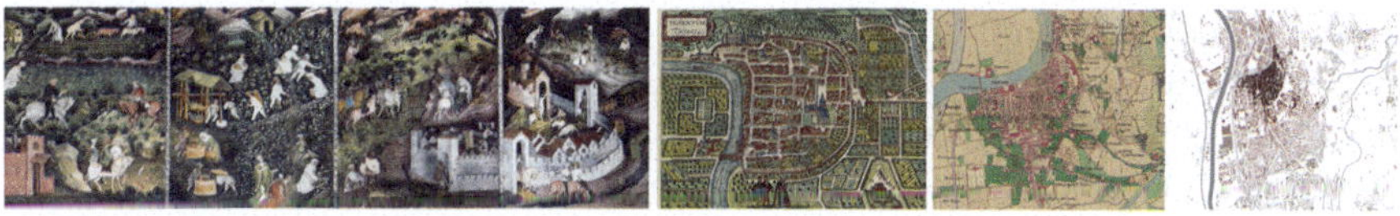

Fig. 8.4 City-countryside relationship in Trento. Late fourteenth century-early fifteenth century, attributed to Master Wenceslas, Cycle of the Months. 1568, Georg Braun, Franz Hogenberg, Tridentum Trient, 53x41 cm, State Archives of Trent, Fondo Carte e Piante. 1812, Napoleonic cadastral map of Trento, Trento Municipal Librar. 2019, Urban expansion from the 1960s onwards. Graphic elaboration by Chiara Frungillo and Giulia Lavolpicella, 2019

towards smaller, often family-run, agricultural operations. Since the early 2010s, there has been a marked shift towards organic farming in Trento, indicating a growing commitment to sustainable agricultural practices. This transition aligns with broader regional initiatives aimed at revitalizing underutilized farmland and promoting sustainable land management. Educational initiatives such as community gardens, farm visits, and school programs are integral to this shift, fostering a deeper connection between the community and local agriculture. These efforts not only contribute to the region's food security but also reinforce Trento's commitment to food system sustainable development.

One of the first projects regarding the local food system was the Nutrire Trento project, initiated in 2017, and still ongoing, in the city of Trento from the collaboration between the University and the Municipality, reflecting a broader trend of local food policy development (Forno et al. 2020). This project aimed to promote conscious consumption, encourage sustainable production, and bridge the gap between producers and consumers. A significant aspect of Nutrire Trento is its monthly roundtable discussions, which facilitate the planning and implementation of actions towards these objectives. Additionally, the project developed a digital platform to map and connect actors within the Trentino short supply chain, creating new opportunities for collaboration. Building on Nutrire Trento's foundation, the HelpFood project in 2021 focused on supporting the purchasing platform and communication activities of the Naturalmente in Trentino Community Supported Agriculture (CSA), as alternative food networks develop diverse business models (Van der Schans et al. 2016). This initiative emerged from the Nutrire Trento #Fase2, which adapted to the changing purchasing habits during the pandemic and developed a platform for local food products. In 2022, the HelpFood4.0 project, funded by EIT Food, was launched as an expansion of HelpFood. Spanning from March to December, HelpFood4.0 aimed to scale up and replicate the socio-technical innovations from its predecessor in other RIS countries. Collaborating with organizations across five EU countries, including the University of Trento, this project sought to enhance local food production by fostering robust food ecosystems. It targeted key areas such as developing livable peri-urban countryside, job creation, promoting biodiversity, and improving the nutritional value of vegetables, all while focusing on social, economic, and ecological sustainability. HelpFood4.0 also involved various Alternative Food Networks in Trentino (Fig. 8.5). These networks faced challenges like inflexible schedules and locations, limited space for social interaction, and issues with product traceability and communication. Building on these initiatives, the Seeds4Future project emerged as a collaborative response to the challenges of urban agriculture, aiming to integrate public and private interests in managing urban parks and abandoned lands for sustainable local food production. This project emphasized the importance of public urban gardens for social interaction and self-production. Seeds4Future aims to enhance urban green management, produce local zero-mile products, and contribute to the EU's sustainability goals. It tests its scalability in the Trentino region and plans to establish at least three living labs in different countries to share and monitor common approaches, promoting

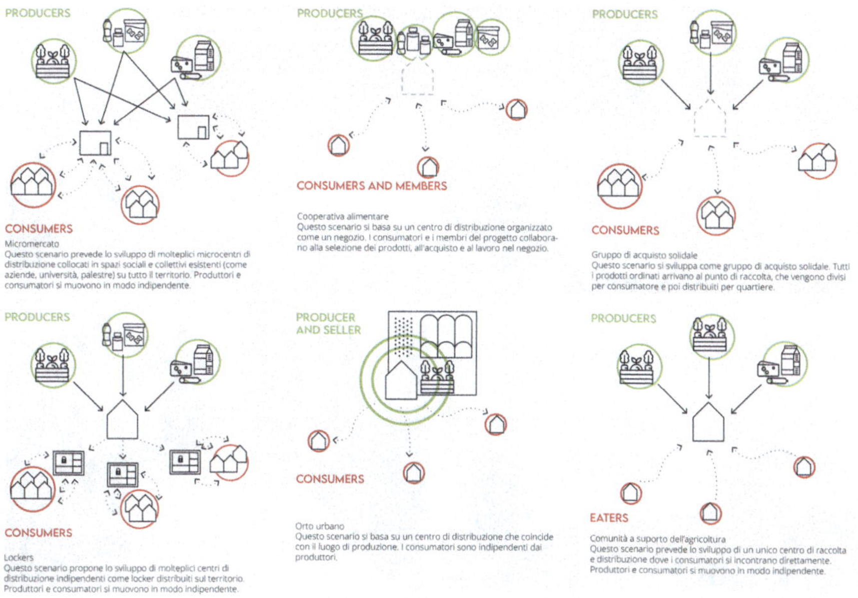

Fig. 8.5 This figure shows stakeholders involved in sustainable food systems within the HelpFood 4.0 project. Graphic elaboration by Giulia Zantedeschi, 2022

urban agriculture as a key element of ecological, economic, and social transformation (Fig. 8.6).

Our research activities commenced with HelpFood4.0 and is carrying on with Seeds4Future, with focus on examining the impacts of food landscapes, particularly urban agriculture, on the territory (Fig. 8.7). We are particularly focused on the ecosystem services provided by agriculture, emphasizing sustainable practices. In our study, we employed a geospatial analysis approach using the Google Earth Engine platform to quantify and assess the ecosystem services provided by urban and peri-urban green spaces in the municipality of Trento. Our ongoing research has made first results leading to preliminary findings. One of the central elements of our analysis is the examination of Surface Urban Heat Island (SUHI) intensity by comparing the Mean Surface Temperature (MST) of urban and rural areas (Fig. 8.8). To conduct this analysis, we have been utilizing data from the Landsat 8 satellite, specifically targeting the data range from July 1st to August 31st across the years 2013 to 2022. This period selection is crucial for capturing the peak summer temperatures, which significantly influence the SUHI effect. We have filtered the satellite data to include images with a cloud cover of less than 20%, ensuring the accuracy and clarity of our observations.

The second result regards the analysis of the carbon storage capacities of various land uses (Fig. 8.9), including agricultural areas, urban parks, and urban forests. The focus was on the calculation of the Normalized Difference Vegetation Index (NDVI) and the estimation of biomass and carbon storage, leveraging high-resolution

Fig. 8.6 This figure shows stakeholder involvement during the workshops' activities in Arco and Trento conducted during the EIT Food Seeds4Future project. Ph. Giulia Zantedeschi, 2024

Fig. 8.7 This figure shows community involvement activities conducted during the EIT Food HelpFood 4.0 project. Ph. Giulia Zantedeschi, 2022

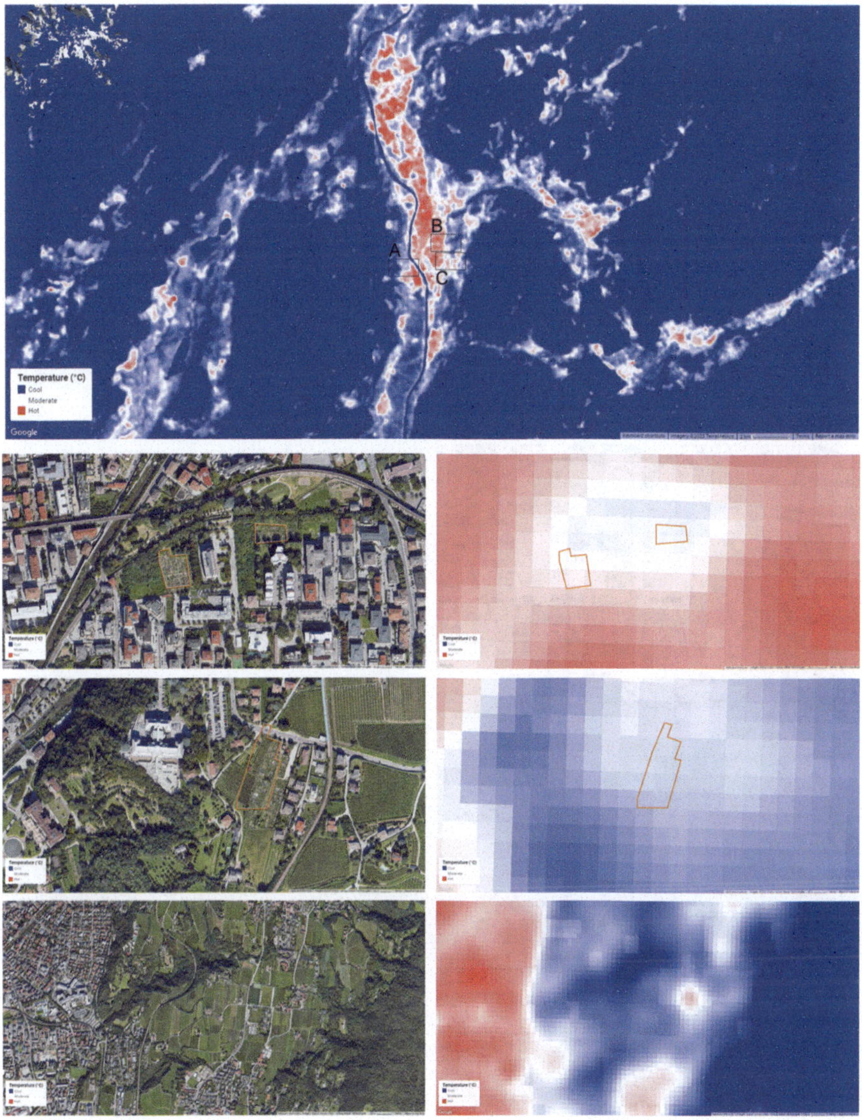

Fig. 8.8 This figure shows the Mean Surface Temperature in cultivated areas. Angelica Pianegonda, 2023

satellite data from the Sentinel-2 mission. The analysis was confined to the summer months (June to August 2023) to capture peak vegetation activity, with a specific criterion set to exclude imagery with cloud cover exceeding 20% to ensure data accuracy and reliability. NDVI, a widely recognized indicator of vegetation health, was computed using the Near Infrared (B8) and Red (B4) bands of Sentinel-2 imagery. Subsequently, biomass estimates were transformed to assess carbon storage, if

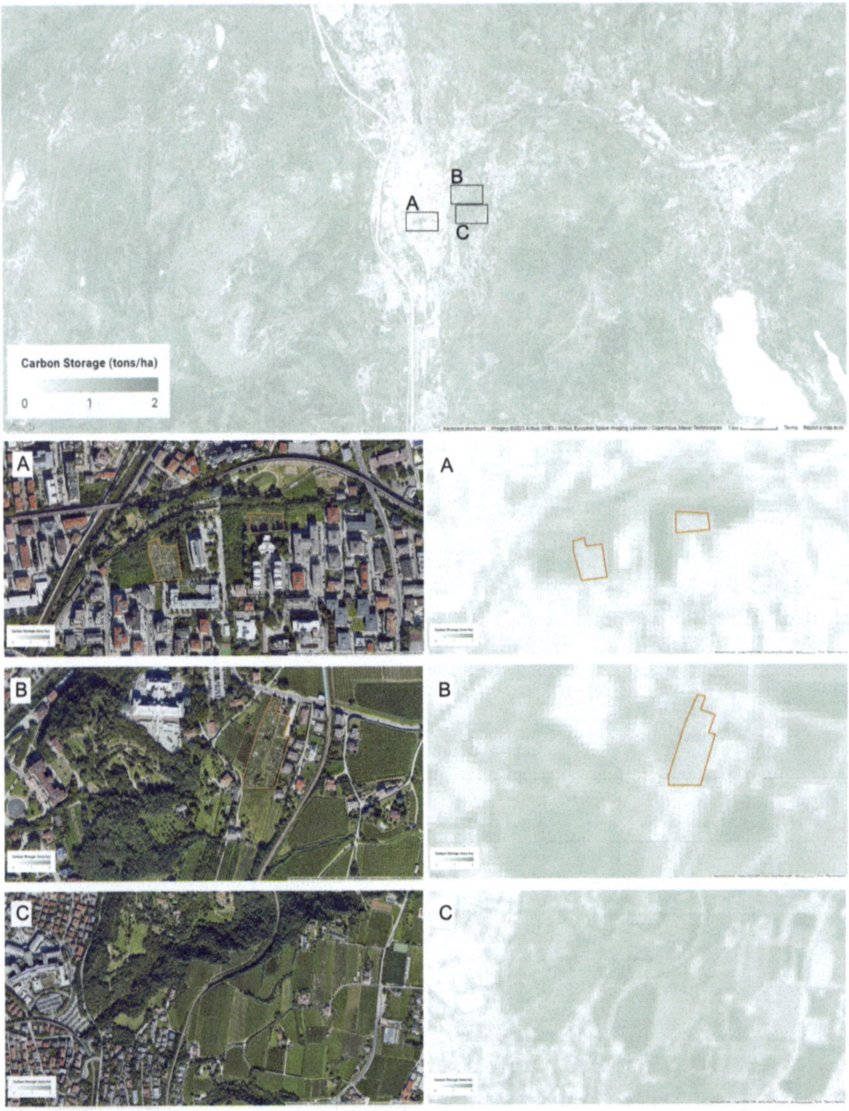

Fig. 8.9 This figure shows the amount of carbon stored in cultivated areas. Angelica Pianegonda, 2023

approximately 50% of the biomass is carbon, a standard approximation in ecological studies. The integration of these datasets facilitated a comprehensive spatial analysis of carbon storage across different land uses within Trento, encompassing agricultural areas, urban parks, and urban forests. This approach not only provided insights into the ecological contributions of these land uses but also enhanced our understanding of the urban-rural gradient in terms of ecosystem service provision.

The visualization of NDVI, biomass, and carbon storage was achieved through the application of color-coded layers on an interactive map, with a carefully designed legend to aid in data interpretation.

These analyses enable us to delve into the environmental benefits and challenges posed by urban agriculture and other green spaces in urban contexts. By understanding the dynamics of the SUHI effect and the ecological contributions of different land uses, with a focus on agriculture, our work aims to inform future urban planning and environmental policymaking, guiding cities towards more sustainable and ecologically balanced futures.

8.4 Conclusion

The food system plays a key role in the process of reconnecting urban and rural landscapes, becoming a driver of innovation and social inclusion, and spreading awareness about food safety and the protection of the natural ecosystem. Agriculture serves as a territorial presidium by decreasing climate change risks, preventing soil erosion, and increasing soil nutrients and biodiversity. Multi-functionality is an intrinsic feature, allowing the development of both productive-material and cultural-social aspects. Good design and management of landscape structures, such as hedges, watercourses, bridges, small forests, villages, orchards, chapels, borders, etc., to establish landscape eco-symbols (Donadieu et al. 2013). The possibility of establishing a symbiotic link between the city and the urban and rural natural environment can be achieved through a new participatory approach to landscape planning that must provide for joint planning between urban and agricultural policies, involving urban, peri-urban, and rural agricultural spaces and the main actors in the landscape project: farmers. The elaboration of the landscape project must be based on the regulation of both public, collective, and private initiatives through integrated planning of the different components. It concerns the recognition of those cultural, symbolic, and formal elements that characterize a certain landscape. In this framework, the agroecological cultivated areas could act as eco-symbols (Donadieu et al. 2013), forming a certain shared vision of a territory.

Sustainable agricultural practices are becoming increasingly widespread, as is the corresponding demand for organic, local, and fair-trade products. This trend affects the whole food system. In this chapter, we focus on the effects it has on the production process and thus on the landscapes of food production, the so-called foodscapes (MacKendrick 2014; Johnston and Baumann 2014; Favargiotti 2019; Favargiotti and Pianegonda 2021). The involvement of citizens and institutions, through projects aimed at understanding and disseminating good practices in the agri-food sector along the entire supply chain, allows us to give further support to these dynamics and act towards a paradigm shift.

Acknowledgements The theoretical approach, scientific position, and conclusions of this contribution have been collectively elaborated by the two authors, Sara Favargiotti and Angelica

Pianegonda. Paragraphs can be attributed to the lead writer as follows: 9.1 Introduction and 9.3 The City of Trento: a case study to understand the impact of the food system on the alpine landscape—Angelica Pianegonda, 9.2 Regenerative foodscapes to restore, regenerate, and reinhabit the existing—Sara Favargiotti, 9.4 Conclusion Sara Favargiotti and Angelica Pianegonda.

The authors acknowledge funding from the EIT Food in the frame of "HelpFood 4.0 - Food Ecosystem Scalability. Assessing digital and cultural potential of Food Hubs towards sustainable local food systems" within the Proof-of-Concept program 2021–2022 (www.eitfood.eu/projects/helpfood4-0) and the "Seeds4future" project within the Proof-of-Concept program 2023–2024 (https://www.eitfood.eu/projects/poc24) lead and coordinated by Sara Favargiotti as Principal Investigator and Scientific Coordinator.

References

Ahn S, Norwood FB (2021) Measuring food insecurity during the COVID-19 pandemic of spring 2020. Appl Econ Perspect Policy 43(1):162–168. https://doi.org/10.1002/aepp.13069

Andreola M, Pianegonda A, Favargiotti S, Forno F (2021) Urban food strategy in the making: context, conventions and contestations. Agriculture 11(2):177. https://doi.org/10.3390/agriculture11020177

Barkema A, Drabenstott M, Cook ML (1993) The industrialization of the US food system. No. 1946-2017-5893

Bocchi S (2020) Agroecology: relocalizing agriculture accordingly to places. In: Bocco S (ed) Bioregional planning and design, vol II. Springer, Cham, pp 81–99. https://doi.org/10.1007/978-3-030-46083-9_5

Branduini PN, Scazzosi L, Laviscio R, Jan S, & Attila TOTH (2016) Urban agriculture and cultural heritage: an historical and spatial relationship. In Urban Agriculture Europe, Jovis, pp. 138–147

Brown KH, Jameton AL (2000) Public health implications of urban agriculture. J Public Health Policy 21(1):20–39. https://doi.org/10.2307/3343472

Cattivelli V, Rusciano V (2020) Social innovation and food provisioning during COVID-19: the case of urban–rural initiatives in the province of Naples. Sustain For 12(11):4444. https://doi.org/10.3390/su12114444

Chambers R (1994) The origins and practice of participatory rural appraisal. World Dev 22(7):953–969. https://doi.org/10.1016/0305-750X(94)90141-4

Colding J, Barthel S, Bendt P, Snep R, Van der Knaap W, Ernstson H (2013) Urban green commons: insights on urban common property systems. Glob Environ Chang 23(5):1039–1051. https://doi.org/10.1016/j.gloenvcha.2013.05.006

De Angelis M (2012) Crises, movements, and commons. Borderlands 11(2):1–22

Donadieu P, De Stefano P, Mininni M (2013) Campagne urbane: una nuova proposta di paesaggio della città. Donzelli

Donovan J, Larsen K, McWhinnie J (2011) Food-sensitive planning and urban design: a conceptual framework for achieving a sustainable and healthy food system. Report commissioned by the National Heart Foundation of Australia (Victorian Division)

Espitia A, Rocha N, Ruta M (2020) Covid-19 and food protectionism: the impact of the pandemic and export restrictions on world food markets. The World Bank https://doi.org/10.1596/1813-9450-9253

Esquivel-Marín NH, Sagarnaga-Villegas LM, Barrera-Perales OT, Leos-Rodríguez JA, Salas-González JM (2023) Multifunctional agriculture in the framework of the sustainable development goals (SDGs): bibliometric review. Acta Univ Sapient Agric Environ 15(1):36–51

Favargiotti S (2019) The foodscape: landscape resources for urban metabolism, social empowerment, and cultural production. Abitare la Terra 50(3):106–107

Favargiotti S, Pianegonda A (2021) The foodscape as ecological system: landscape resources for R-urban metabolism, social empowerment and cultural production. In: Catalano C, Andreucci MB, Guarino R, Bretzel F, Leone M, Pasta S (eds) Urban services to ecosystems: green infrastructure benefits from the landscape to the urban scale. Springer, Cham, pp 279–295. https://doi.org/10.1007/978-3-030-75929-2_15

Favargiotti S, Ferretti M, Elzenbaumer B, Dalla Torre C (2023) RURAL COMMONS FESTIVAL. Pratiche emergenti di comunità per la cura e gestione dei beni comuni rurali. In: Morozzo della Rocca MC, Olivastri C, Zappia G, Zignego MI (eds) Design, comunità e aree rurali. Sinergie e prospettive. De Ferrari Editore, Genova, pp 59–70

Ferretti M, Favargiotti S (2022) Commons in marginal landscapes: collective practices for an alternative narrative and use of common spatial resources in peripheral landscapes. Ri-Vista Res Lands Archit 19(2):176–189. https://doi.org/10.36253/rv-11412

Fitzpatrick K, Harris C, Drawve G (2020) Assessing US food insecurity in the United States during covid-19 pandemic. Montana 34(3):3. https://www.cresp.udel.edu/wp-content/uploads/2020/05/Assessing-Food-Insecurity-Brief.pdf

Food and Agriculture Organization of the United Nations (2011) The state of food insecurity in the world 2011. Rome. http://www.fao.org/3/i2330e/i2330e00.htm

Forno F, Maurano S, Vittori F (2020) Costruire processi partecipativi attorno al cibo: Le esperienze di Bergamo e Trento. In: Dansero E, Marino D, Mazzocchi G, Nicolarea Y (eds) Lo Spazio delle Politiche Locali del Cibo: Temi, Esperienze e Prospettive. Celid, Torino, pp 15–30

Fung A (2015) Putting the public back into governance: The challenges of citizen participation and its future. Public administration review, 75(4), 513–522

Gaitán-Cremaschi D, Klerkx L, Duncan J, Trienekens JH, Huenchuleo C, Dogliotti S, Rossing WA (2019) Characterizing diversity of food systems in view of sustainability transitions. A review. Agron Sustain Dev 39(1):1–22

Giudici G (2021) Urban relations and dematerialization of production processes. How the Covid-19 epidemic has changed our society and our cities

Gliessman S (2018) The co-creation of agroecological knowledge. In: Gliessman S (ed) Agroecology: the ecology of sustainable food systems. CRC Press, Boca Raton, pp 23–47

Harvey D (2012) Rebel cities: from the right to the city to the urban revolution. Verso Books, London

Hjalager AM, Richards G (2003) Tourism and gastronomy. Routledge, Milton Park

Johnston J, Baumann S (2014) Foodies: democracy and distinction in the gourmet foodscape. Routledge, Milton Park

Keil MML, Kistmann VB (2016) The sense of belonging: emotional aspects in participatory design in relation to products and services for cities. Strateg Design Res J 9(3):163–171

Lovell ST (2010) Multifunctional urban agriculture for sustainable land use planning in the United States. Sustain For 2(8):2499–2522. https://doi.org/10.3390/su2082499

Luck R (2018) Participatory design in architectural practice: changing practices in future making in uncertain times. Des Stud 59:139–157. https://doi.org/10.1016/j.destud.2018.10.003

MacKendrick N (2014) Foodscape Contexts 13(3):16–18. https://doi.org/10.1177/1536504214545754

Marot S (2019) Taking the Country's side: agriculture and architecture. Lisbon Architecture Triennale

Méndez VE, Bacon CM, Cohen R (2013) Agroecology as a transdisciplinary, participatory, and action-oriented approach. Agroecol Sustain Food Syst 37(1):3–18

Moreira S, Morell MF (2020) Food networks as urban commons: case study of a Portuguese "prosumers" group. Ecol Econ 177:106777. https://doi.org/10.1016/j.ecolecon.2020.106777

Nathan M, Overman H (2020) Will coronavirus cause a big city exodus? vol 47. Econ Observ, pp 1537–1542

Opitz I, Berges R, Piorr A, Krikser T (2016) Contributing to food security in urban areas: differences between urban agriculture and peri-urban agriculture in the Global North. Agric Hum Values 33(2):341–358. https://doi.org/10.1007/s10460-015-9610-2

Pelenc J, Wallenborn G, Milanesi J, Sébastien L, Vastenaekels J, Lajarthe F, Frère B, Cervera-Marzal M, Carimentrand A, Merveille N, Frere B (2019) Alternative and resistance movements: the two faces of sustainability transformations? Ecol Econ 161:1–10

Rosset P (2008) Food sovereignty and the contemporary food crisis. Development 51(4):460–463. https://doi.org/10.1057/dev.2008.48

Sanyé-Mengual E, Specht K, Krikser T, Vanni C, Pennisi G, Orsini F, Gianquinto GP (2018) Social acceptance and perceived ecosystem services of urban agriculture in Southern Europe: the case of Bologna, Italy. PLoS One 13(9):e0200993. https://doi.org/10.1371/journal.pone.0200993

Snapp S (2017) Agroecology: principles and practice. In: Agricultural systems. Academic, London, pp 33–72

Steel C (2013) Hungry city: how food shapes our lives. Random House, London

Sumner J (2011) Serving social justice: the role of the commons in sustainable food systems. Stud Soc Just 5(1):63–75

United Nations, Department of Economic and Social Affairs, Population Division (2017) World Population Ageing 2017 (ST/ESA/SER. A/408)

Van der Schans JW, Lorleberg W, Alfranca O, Alves E, Anderson G, Branduini PN, Egloff LD, Giacchè G, Heller A, Herkströter K, Kemper D, Koleva G, Mendes Moreira P, Miguel A, Neves L, Paulen O, Pickard D, Padros MJ, Pölling B, Recasens X, Ronchi B, Spornberger A, Timpe A, Torquati B, Weissinger H, Wydler H (2016) It is a business! Business models in urban agriculture. In: Urban Agriculture Europe. Jovis, pp 82–91

Vrublova K (2020) Evaluation of ecosystem services loss due to urban sprawl on agricultural land in the context of sustainable development. J Lands Ecol 13(3):122–133

White R (2007) Green criminology and the pursuit of social and ecological justice. Issue Green Criminol:32–54

Zasada I (2011) Multifunctional peri-urban agriculture—a review of societal demands and the provision of goods and services by farming. Land Use Policy 28(4):639–648

Chapter 9
Growing Urban Health: Questioning the Role of Urban Gardening in Distressed Urban Areas

Livia Calcagni and Alberto Calenzo

Abstract Given its integrated benefits, urban agriculture, especially community gardens, may significantly enhance the socio-environmental condition of vulnerable districts and communities. This chapter examines different aspects of urban agriculture to highlight its value beyond profitability and food production in terms of social empowerment, health, and well-being. The chapter seeks to understand the reasons behind the failure of a community garden initiative that took place in a distressed urban area of the city of Rome and to clarify the connections between community garden initiatives, socioeconomic context, and urban structure. Multiple-case study research was conducted to compare successful best practices with the case study mentioned above to identify invariants and draw cross-case conclusions. All case studies are linked to bottom-up processes in marginal areas where public space could represent a catalyst capable of embracing cultural, social, environmental, and eco-systemic matters. The study indicates how critical elements for a successful and long-lasting implementation of community garden initiatives must include identification of the community's primary needs, institutional and financial support, extensive engagement of inhabitants, and the role and diversification of stakeholders. These aspects drive the meta-design phase and long-term planning of the overall process.

Keywords Urban gardening · Urban health · Distressed urban areas · Case study research

L. Calcagni (✉)
Department of Planning, Design and Technology of Architecture (PDTA), Faculty of Architecture, Sapienza University of Rome, Rome, Italy

Department of Digital Architecture and Planning, Faculty of Architecture and Planning, Vienna University of Technology, Vienna, Austria
e-mail: livia.calcagni@uniroma1.it

A. Calenzo
Department of Planning, Design and Technology of Architecture (PDTA), Faculty of Architecture, Sapienza University of Rome, Rome, Italy
e-mail: alberto.calenzo@uniroma1.it

A. Battisti et al. (eds.), *Urban-Rural Dialogue in Green City Design*, Designing Environments, https://doi.org/10.1007/978-3-031-88318-7_9

9.1 Introduction

The work presented in this chapter is part of a broader research-action project that is characterized by a community-based approach. The project is led by a multidisciplinary research group of Sapienza University of Rome in collaboration with other social, health, and local institutions, including the Local Health Unit (ASL Roma 1), the Department of Epidemiology of the Health Regional System of Lazio (DEP), the XIII Municipality of Rome, the National Institute for the Promotion of the Health of Migrant Populations, and the Contrast of Diseases of Poverty (INMP). The research aims to investigate the effect of the unequal distribution of social determinants on the health of a population characterized by poor housing and socioeconomic vulnerability. More specifically, the research intends to develop regeneration strategies for the distressed urban area of former Bastogi complex in the northwest of Rome, Italy. The research group has focused on the conditions of neglect that characterize outdoor public spaces in the complex. Within this context, an urban farming initiative led by the neighbourhood Committee took place. This urban agriculture (UA) initiative, launched by the Committee at its participatory peak, sadly ended after only 2 years of practice. This chapter examines the reasons behind the initiative's failure and seeks to clarify the connections between urban agriculture, socioeconomic context, and urban spatial patterns.

According to a group of experts of the Instituto Universitario de Urbanística de la Universidad de Valladolid integrated urban regeneration is considered as the natural outcome of joining together two basic recommendations of the Leipzig Charter: making greater use of integrated urban development policies and paying particular attention to deprived neighbourhoods (Alvarez Mora and Roch Peña 2010). Urban regeneration focuses primarily on building and physical upgrading, such as improving the energy efficiency of the residential stock, without considering the role it can play in enhancing inhabitants' social and economic conditions. Urban agriculture, especially in distressed urban areas, can blur the lines between the natural, the artificial, and the cultivated, providing a linked network of natural and semi-natural elements capable of providing multiple functions and ecosystem services with positive economic and social benefits for humans and other species (Battisti and Calcagni 2023; Naumann et al. 2011; Benedict and McMahon 2012). The Food and Agriculture Organization and several other international organizations have promoted initiatives and released reports on implementing efficient and sustainable food production systems to ensure access to quality food and liveable environments. According to the COST Action on Urban Agriculture Europe (Lohrberg et al. 2016), urban agriculture spans all actors, communities, activities, places, and economies focusing on biological production in an 'urban' context. It is structurally embedded in the urban fabric and integrated into the city's social and cultural life, economics, and metabolism. Within the broader urban agriculture category, it is possible to outline a more precise subcategory: urban community gardening. Community gardening comprises any piece of land (publicly or privately held) that is cultivated by a group of people based on bottom-up initiatives and tended collectively. In this case, the predominant purpose is not food production but social functions for community empowerment. This is why this subcategory is suitable for distressed urban areas, which, according to the OECD definition, are situations of underdevelopment in developed contexts, that is, areas of a city that suffer social, economic, cultural, and ecological deprivations within the city. Distressed areas have higher concentration of low-income households, suffer from physical deterioration of infrastructure, higher crime rates and vandalism and other similar socio-economic deprivations (OECD 1998). As a nature-based solution and green infrastructure, UA brings about ecological, health, economic, and social integrated benefits (Fig. 9.1), creating a direct link between producer and consumer, giving birth to the so-called prosumer.

In terms of mental health and sense of well-being, for instance, community gardens have a huge impact (Ilieva et al. 2022; McGuire et al. 2022):

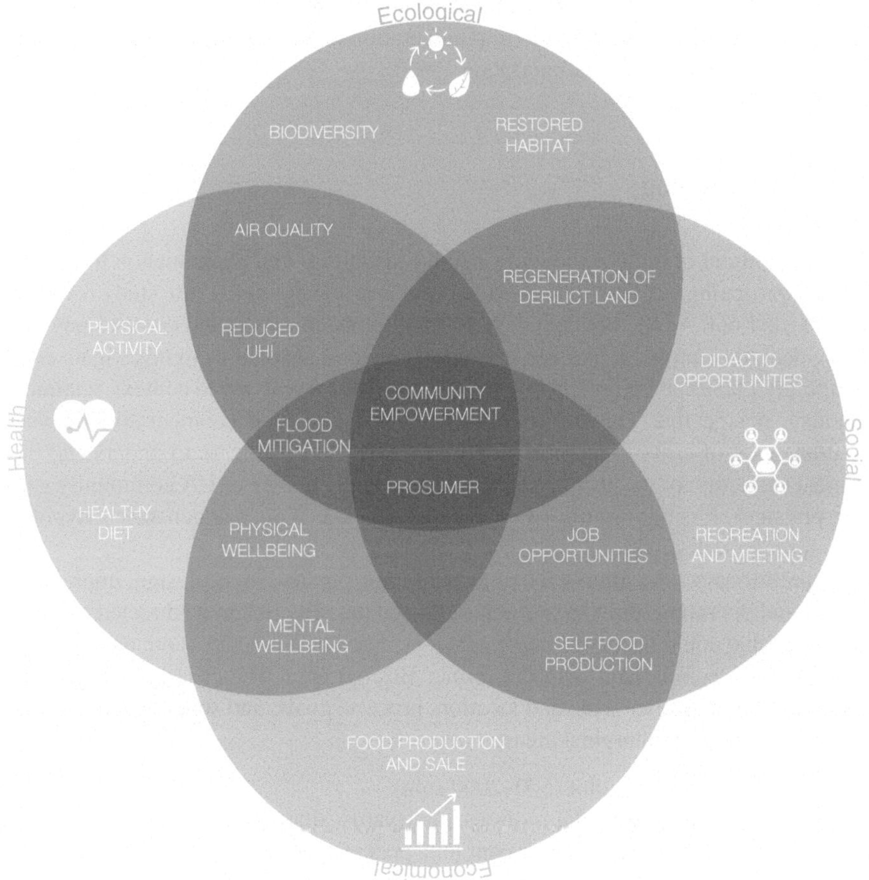

Fig. 9.1 Integrated and overlapping social, economic, ecological and health benefits of community gardens as green infrastructures

- promotion of a healthier lifestyle by fostering outdoor physical activity and a correct food diet
- improvement of air quality by absorbing atmospheric CO_2
- reduction of urban heat island effect and mitigation of microclimate conditions
- community empowerment and engagement, offering inhabitants meeting and social opportunities
- promotion of the biophilia effect
- creation of enhanced recreation spaces and indirect education and didactical opportunities on sustainable and quality food issues

However, the lack of formal recognition of urban agriculture in planning policy, the lack of awareness about the socio-economic and environmental role of UA in cities, the lack of clear government responsibilities for the management of UA, land use issues - specifically availability of land, access to land and usability of land - and the lack of resources, technical and financial support from the government for UA, pose significant hurdles not only to the implementation of UA but especially to its long-term management and operation (Quon 1999).

9.2 Methodology

Given the intent of understanding in-depth a specific social phenomenon like community gardening, seeking to understand why and how it works, case study research was carried out. As the boundaries between the phenomenon and context may not be evident, the empirical case study research method enabled us to investigate contemporary phenomena (projects) in depth and within a real-world context, retaining a holistic perspective. The review covered multiple cases and a comparative analysis to draw a set of cross-case conclusions (Yin 2018), allowing us to understand the critical elements for the success, acceptance, and popularity of UA community garden practices. It is both illustrative and exploratory, as it aims at generating hypotheses for later investigation.

A preliminary screening of the objectives to be pursued in the design, implementation, and operation of the case studies allowed the selection to be directed towards examples that could be, on the one hand, considered best practices and, on the other, suitable for a comparison with the former Bastogi area. The criteria addressed in their selection consider scale and location, process, goals, and stakeholders.

Building scale in marginal areas:

- small to medium size plot (500–2000 sqm)
- shared building spaces: flat rooftops, courtyards, shared gardens
- unused/derelict urban space: vacant lots, pocket spaces

Participatory processes and bottom-up initiatives:

- community needs and requirements
- support from local authority/municipality / public administration
- participation in all phases

Project objectives:

- local food production for self-consumption
- environmental sustainability
- public shared spaces for community involvement
- educational, recreational, and rehabilitator activities

Stakeholders:

- neighbourhood inhabitants (0–100 aged)
- architect, landscape architect, technical figures
- municipality, public administration
- local associations, committees, voluntaries

To construct validity (Yin 2018) the case study analysis relied on multiple sources of evidence, ranging from scientific and grey literature, document analysis (municipality reports and administrative documents), media coverage (online newspaper articles, architecture websites), and direct observations. To prevent biased documentation, a process of corroboration of the information allowed the comparison of different sources (Yin 2018). The data for each case study was analysed by applying an assessment protocol developed in previous studies by the authors (Calcagni and Calenzo 2023) to identify invariant patterns and issues. In addition to a whole series of data relating to the project (year of construction, location, type of project, main characteristics, actors involved), the development over time of the entire process (design, construction, and operation) was analysed to detect the key factors that contributed to the success or failure of the initiative. Finally, each case study was evaluated according to three main criteria and sub-criteria: social impact (community empowerment, related activities, social catalyst), management (participatory process, administration, funding), and cost-effectiveness (low-tech techniques, recycling, and reuse, food self-production). For each sub-component, a score was assigned to obtain a global one for each case study, thanks to their sum.

9.3 Urban Gardening Initiatives Across Europe: 5 Case Studies

Case 1: De Ceuvel, Amsterdam, The Netherlands (Fig. 9.2). This project focuses on environmental and social regeneration of a former shipyard in the industrial area of Buiksloterham, Amsterdam North, once heavily polluted, into an innovative, sustainable community and innovation hub. In 2012, the land was secured for a 10-year lease from the Municipality of Amsterdam after a local thriving community of young entrepreneurs and artists, supported by a group of architects, won a tender to turn the site into a regenerative urban oasis. The goal was to create a low-impact biomass production area, conceived as a playground for sustainable technologies, as much energy self-sufficient as possible: processing waste and producing food. The

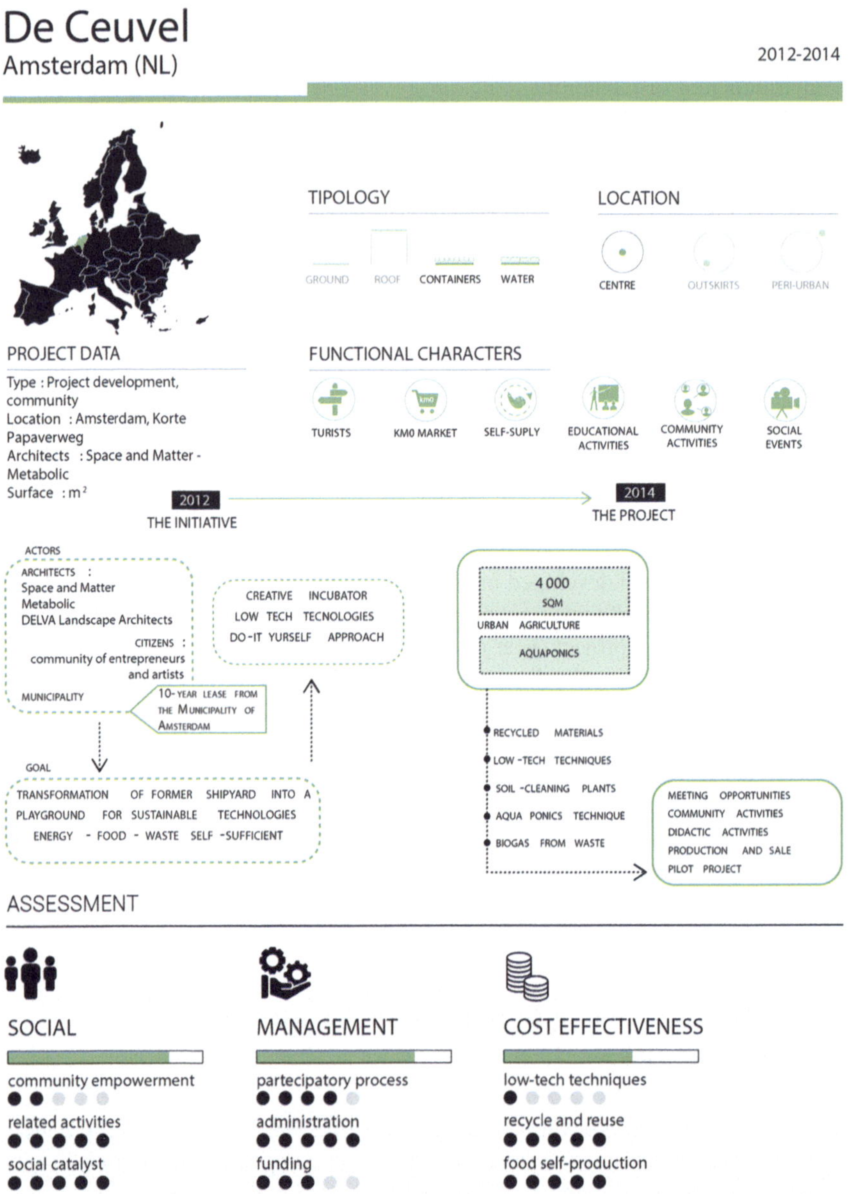

Fig. 9.2 De Ceuvel, Amsterdam (The Netherlands). Main features, process development and analysis

greenhouse is the key to recycling nutrients at De Ceuvel by integrating it with aquaponics. It produces vegetables and herbs for Cafe de Ceuvel using a closed-loop aquaponics system combining fish and vegetable production. The fish excretes are broken down into nutrients for the plants, and the plants provide a natural filter for the water in which the fish live. Inputs include primarily local nutrients like worms from composting bins and struvite from their struvite reactor, produced by human urine from the community. De Ceuvel has proved to be an engaging educational environment for the starting community and an attraction for all the other neighbourhood inhabitants, becoming a new facility for the entire area, still lacking other services and facilities (De Ceuvel; Assemble Papers 2015; Barba Lata and Duineveld 2019).

Case 2: Esto no es un solar 2–27, Zaragoza, Spain (Fig. 9.3). "Esto no es un solar" means "This is not a plot of land," to emphasize that an empty space, mainly used for garbage and crime, within the city, can become a collective space due to temporary use strategies. These spaces are equipped with a minimum cost, and a contract is established with the owner, who gives it for free to the municipal administration. Most of the neighbourhood population was involved in the conception and construction of the projects, as well as unemployed citizens or people involved in socially useful jobs. Within this strategic program, more than 20 spaces underwent a regeneration process between 2006 and 2014. Amongst these two areas were assigned to urban farming. Installing portable structures was conceived to revitalize the area, thus generating the missing urban function. San Pablo neighbourhood hosted the first intervention and an experimental program. All the materials used are low-cost and often recycled. A series of wooden pallet platforms define green spaces of lavender, rosemary, and different types of plants as a botanical garden. It is also a space for didactic activities for local nearby schools and is entirely managed by the adjacent senior centre (as the design phase had established). A second urban garden was planned, and on this occasion, a group of users was also formed (local Public School, Children's centre, Senior Centre), guaranteeing a variety of users. The spaces that had not been entrusted to associations that were responsible for them have been subject to decay and abandonment. An incorrect interpretation of the population's needs has emerged in some cases, and some foreseen uses have yet to be accepted (Esto no es un solar 2020; Archdaily 2014a, b, 2016).

Case 3: The StadtAkcer, Munich, Germany (Fig. 9.4). The StadtAcker is a bottom-up initiative for a community garden of around 1.000 sqm in Munich in the Ackermannbogen district. This initiative came from the Ackermannbogen association, in collaboration with citizens—who strongly wanted the garden—and the Municipality. The idea was born in 2011, but it had to wait around 6 years to be allocated a gardening plot on municipal property. In the meantime, the idea was actively promoted and kept alive among the population through small, mobile, and decentralized gardening events in the neighbourhood. As the garden is intended to be an authentic community garden, the main goal was to establish an open space for sharing knowledge and cooperation and improving social cohesion.

For this reason, there are no single plots, but the entire gardening group maintains the garden together, subdivided into thematic specialized sub-groups such as vegetables, herbs, compost, berries, or bees/flowers. The rules and usage conditions of the StadtAcker community garden were defined through speakers of the different

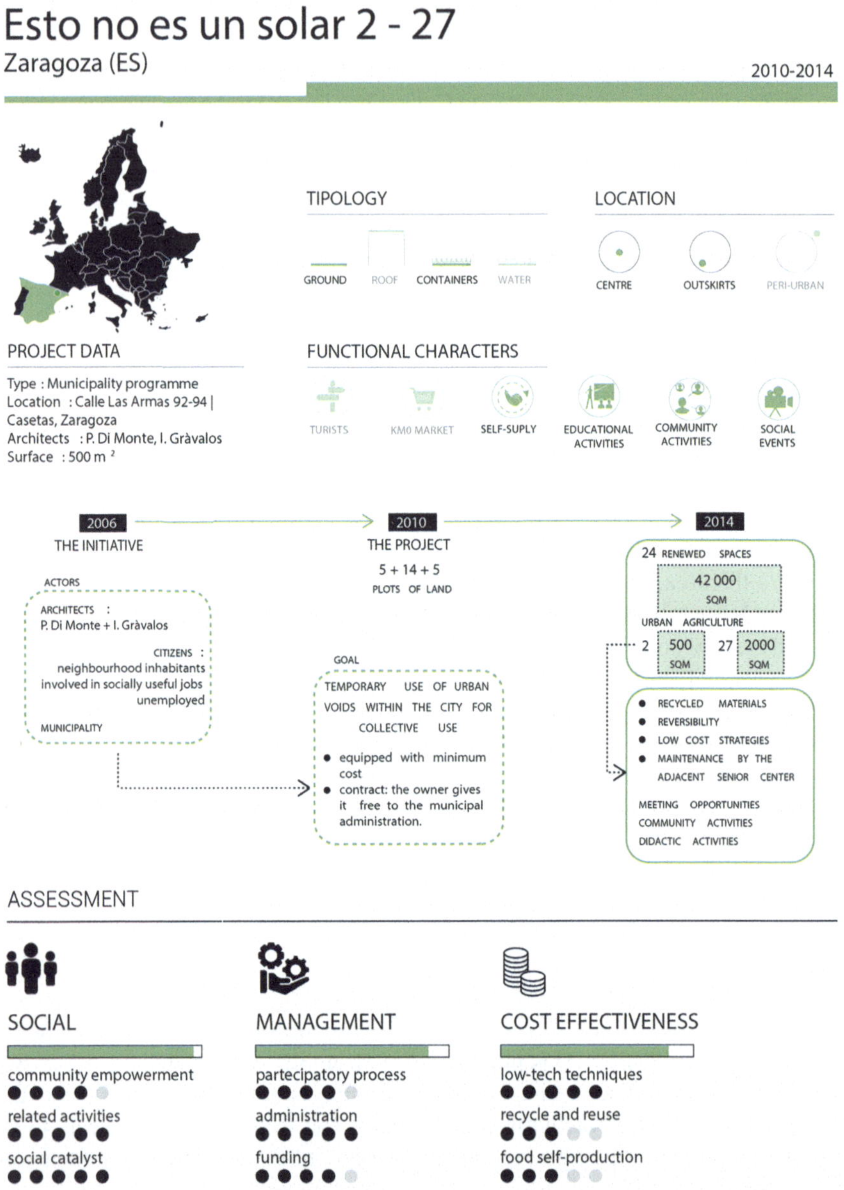

Fig. 9.3 Esto no es un solar, Zaragoza (Spain). Main features, process development and analysis

gardening groups in close cooperation with the Association. The Municipality of Munich supported the realization of the infrastructure of the garden (i.e., fencing, groundwater, storage facility, soil, and other aspects), and each participant contributed an annual fee for its maintenance.

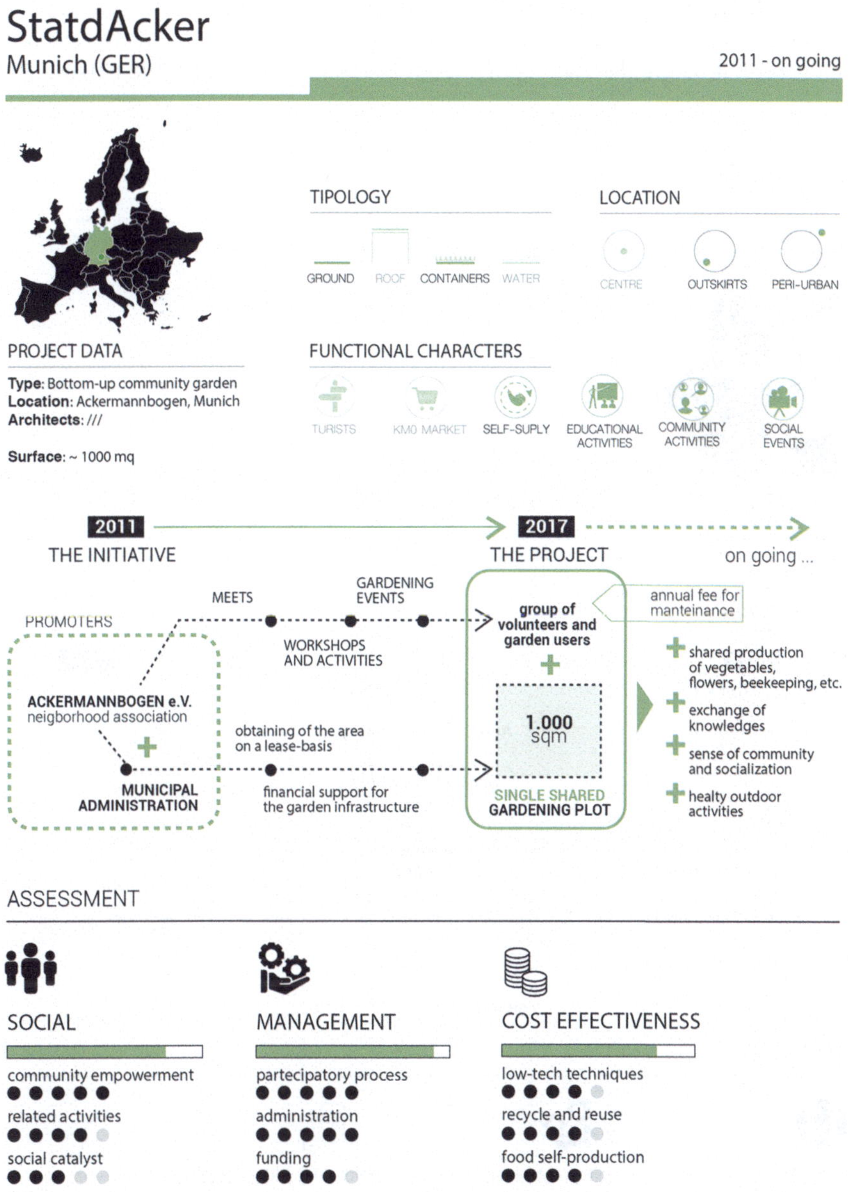

Fig. 9.4 StadAcker, Munich (Germany). Main features, process development and analysis

After the first year of operation, the garden is perceived as a valuable long-term neighbourhood feature. Of course, Municipalities are crucial because they can provide aid and resources—first, the soil—to initiatives that cannot sustain their projects only based on volunteering (Environment & Society Portal 2019; LMU 2020).

Case 4: Agrocitè, Colombes-Gennevilliers, France (Fig. 9.5). Agrocité is a project for an urban agriculture hub designed in 2013 by the "Atelier d'Architecture Autogérée" in Colòmbes as part of their broader project called R-Urban. The

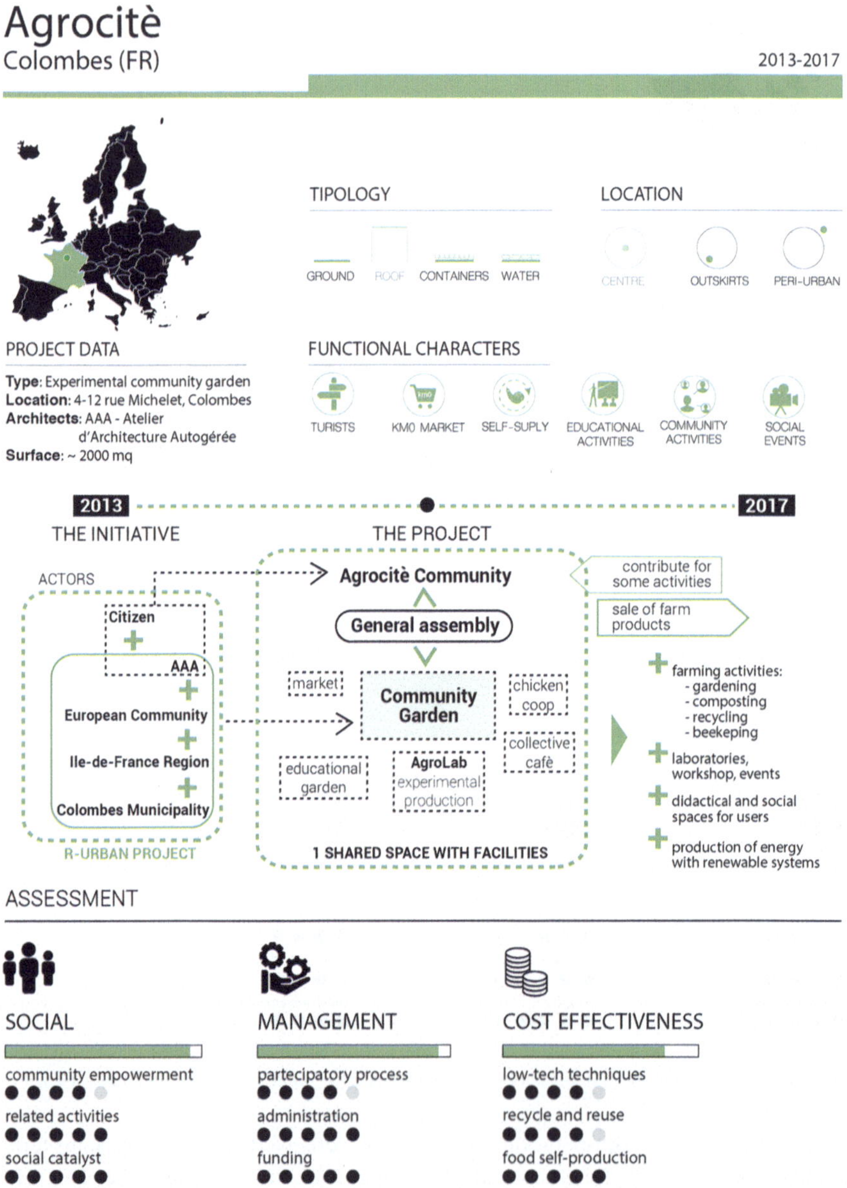

Fig. 9.5 Agrocitè, Colombes, Gennevilliers (France). Main features, process development and analysis

project consists of an area for community gardening with a shared garden for residents, an educational garden, an 'AgroLab' specialized in experiments with intensive production, a shared greenhouse, equipment for rainwater collection and other devices for energy production from renewable sources, a vegetable market, and a Café. Several collective activities—widely sponsored via different media platforms—occur within the garden. The events are related to gardening practices and ecology, as well as to cooking, composting, and recycling. The workshops and laboratories organized are previously discussed and agreed upon through the General Assembly, in which some user representatives and some members of the Architecture collective take part. In 2017, despite numerous petitions and protests from citizens, the initiative was removed to make place for a parking lot promoted by the new local administration. Luckily, the following year, a new edition of Agrocitè found a place in the suburbs of Paris. Although the concept and the project are identical, it will take time to recreate the feeling of belonging and the social cohesion that characterized the first experiment, highlighting how users' participation can be considered the core for the success of this kind of project (Atelier D'Architecture Autogeree; R-Urban 2012; Le Parisien 2017).

Case 5: Student Campus Community Garden, Ljubljana, Slovenia (Fig. 9.6). This case study is in a student community rather than a residential one. In 2015, the student community in Ljubljana reunited as a collective and set off a community garden initiative. This experience is very interesting because, despite being self-managed and self-financed, it has survived until now. In cooperation with the administration of Student Homes in Ljubljana, students managed to acquire land near dormitories based on a self-management agreement. They created two community gardens, one in Rožna Dolina and another in student dormitories in Mestni Log. The gardens are intended primarily for students. However, anyone interested is well accepted. The project's vision is to co-create a sustainable community space dedicated to socializing, connecting, learning, relaxing, and producing food, adding value to the space through various events and activities that encourage the creative use of public spaces. Participants are responsible for the entire garden area, from maintaining the basic infrastructure and equipment, including sheds, rainwater butts, and composters (in cooperation with the canteen kitchens) to landscaping the wider campus by planting herbs, spices, and fruit trees. The division and planning of the work are coordinated during regular weekly meetings. The student community garden project is an example of good cooperation between a civil initiative and a public institution. It is one of the rare examples of a civil initiative that has been maintained and continues to be self-managed. Also, in this case, the Public Administration support was crucial.

Case 6: Very Important Quarter initiative in former Bastogi, Rome, Italy (Fig. 9.7). The former Bastogi-complex is in the northwest of Rome and consists of 6 buildings built in the 80's as a residence for Alitalia employees. After some years, it was assigned by the Municipality of Rome to deal with the housing emergency. The complex was designed to accommodate users for short stays, but today, it has a stable population of more than 2000 inhabitants, according to qualitative studies (Battisti et al. 2021), twice as much as the figures provided in the ISTAT 2011

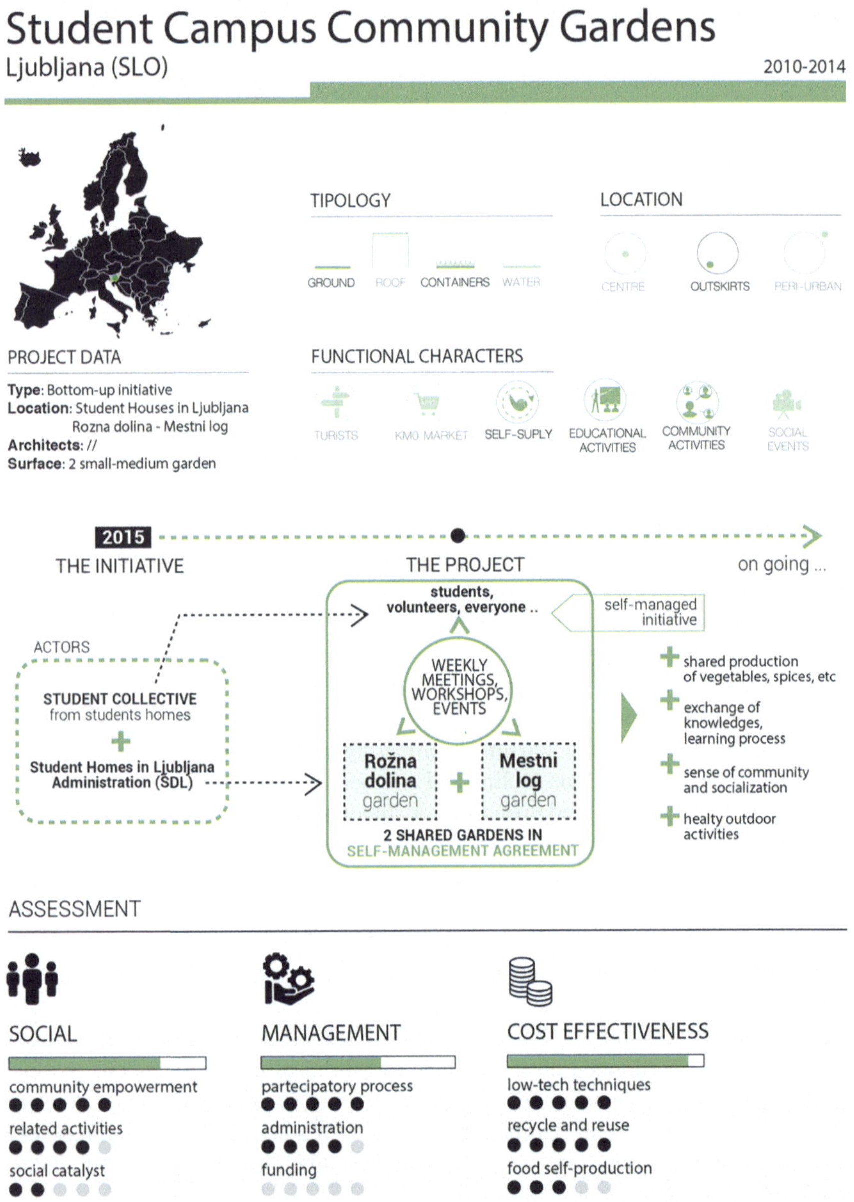

Fig. 9.6 Student Campus Community Gardens, Ljubljana (Slovenia). Main features, process development and analysis

census. The awareness that former Bastogi was a complex system led to a preliminary assessment of the social, environmental, and architectural conditions to achieve an overall intersectoral understanding of the context. Community activities, such as

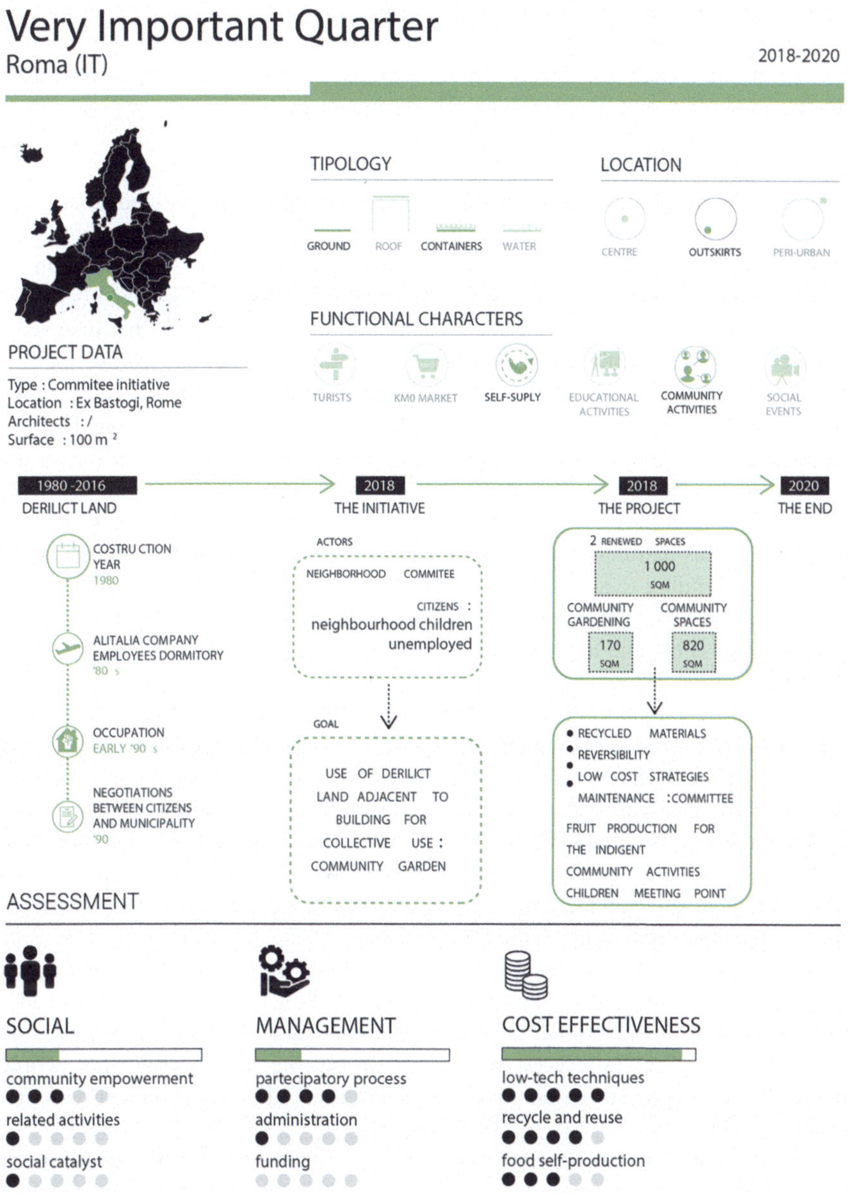

Fig. 9.7 Former Bastogi "Very Important Quarter" initiative, Rome (Italy). Main features, process development and analysis

focus groups, questionnaires, and community consultations with some members of the local population, followed this phase. The combination of qualitative and quantitative analysis showed how the former Bastogi is characterized by spatial and social segregation and marginalization. Former Bastogi is an enclave within the urban fabric of Rome that hosts various types of vulnerable groups: children and older people in a state of neglect, illegal migrants, people under house arrest, drug addicts, alcoholics, and people with HIV (Battisti et al. 2020).

In 2018, the Bastogi neighbourhood Committee launched the urban community gardening initiative. The Committee at that time consisted of approximately ten members, and the plot chosen for the garden was a derelict piece of land adjacent to one of the buildings. The singularity of former Bastogi lies in the impossibility of ascribing it to any legal status in terms of property (Battisti et al. 2021) making it complicated for the inhabitants to undertake practical actions on their territory. This experience was characterized by the inhabitants' appropriation of this dismissed "public" green space without a formal agreement with the Municipality. Nevertheless, this initiative could be considered a bottom-up planning action for shared purposes rather than the expression of urban informality or an unauthorized action. The aim was to create a community garden, which could also represent an alternative for the young residents (children and teenagers), offering a new positive space for free time activities in which they could experience a healthy lifestyle. Considering the tenure's legal status and consequent management implications, the Committee could not find any funding to support the initiative. This led the Committee members to use either recycled or leftover materials and low-tech techniques to limit the budget as much as possible and to create a reversible solution given the uncertainty of the property.

Social and economic benefits included an initial increase in civic engagement from the neighbourhood inhabitants, the promotion of outdoor physical activity, and access to quality fresh food. Social cohesion and community engagement are essential, especially in a vulnerable area characterized by one of Rome's highest social hardship indexes. The Italian PASSI surveillance (Progressi delle Aziende Sanitarie per la Salute in Italia) for the years 2015–2018 reveals that one in 5 deaths is linked to an inadequate diet and that only 10% of the population consumes five portions of fruit or vegetables recommended by the WHO.

PASSI also reports that over 37% of individuals do not reach recommended physical activity levels, and the graphs demonstrate how a healthy diet, and an adequate level of physical activity are inversely proportional to income and education level, therefore potentially very low in former Bastogi. Considering that people spend 22 h a day in indoor environments (U.S. Environmental Protection Agency 2013).

the condition in former Bastogi is further aggravated by the terrible conditions of the housing accommodations, characterized by mould, humidity, and overcrowding (Battisti et al. 2021). All things considered, the community garden offered opportunities for healthy lifestyles, adequate diet, and outdoor physical activity—that largely contribute to the decrease of Non-communicable Diseases (reference)—contributing to the overall health and social well-being of the inhabitants.

Another positive key element that distinguished this initiative was the use of derelict land for social purposes, as well as the intergenerational character of the group, which helped to strengthen, even if minimally, the sense of community and belonging.

In terms of environmental benefits, introducing fruit trees contributed to CO2 absorption, air quality improvement, and reduction of the urban heat island effect.

The Committee, which was entirely responsible for managing the community garden, gradually lost several members because of the absence of a structured organization defining the roles and duties of each component. The outburst of the COVID-19 pandemic in 2020 largely contributed to the lack of interest in participating in the daily activities necessary to maintain the community garden. The informal and short-term planning and the lack of promotion to the community through events, workshops, and reunions failed to involve new members and participants throughout the years. In addition, the lack of a formal agreement with the Municipality and no external funding for resources and training contributed to the members' lack of appropriate equipment and know-how. As a result, the initiative gradually lost momentum and ended in 2021.

9.4 Conclusions

The comparison between the best practices and the case study provides sufficient data to draw some conclusions about the constraints and challenges faced by the former Bastogi case study (Fig. 9.8). First, the urban farming initiative in Bastogi

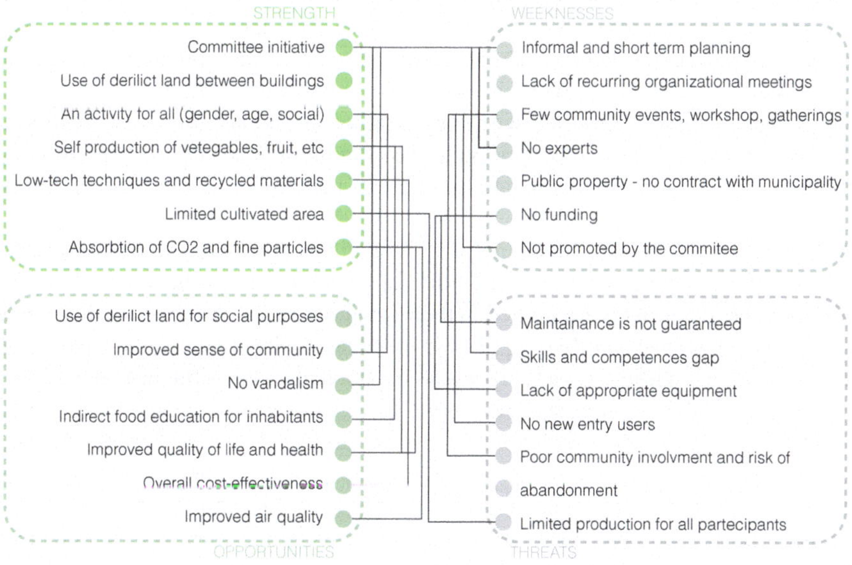

Fig. 9.8 SWOT analysis and correlations between features that distinguished the "Very important quarter" urban gardening initiative in former Bastogi, Rome

has been dramatically affected by the absence of institutional and financial support. The lack of dialogue and support with the local administration, combined with the ignorance of the opportunity to participate in tenders or funds, widely contributed to the short-lasting lifetime of the initiative. Another trigger to the project's failure is the poor promotion and the lack of engagement of the inhabitants on a large scale. More specifically, the number of participants over time not only did not grow but also decreased, presumably due to the lack of social cohesion and general diffidence among inhabitants towards new initiatives and community activities, as also confirmed by the low turnout during the focus groups carried out during the analytical phase (Battisti et al. 2021).

Since the initiative was launched by a relatively small initial community (not bigger than 8–10 members), the communication and promotion of the initiative should have started already in the preliminary design phase to ensure from the very start the engagement of a more significant portion of the community. Moreover, the lack of resources and know-how, combined with the absence of qualified personnel that could either support the community or provide basic technical training, represented a limit to the development and maintenance of the initiative.

Another critical aspect is the absence of a long-term design approach combined with a disorganized and non-structured organization that could not manage, divide, and assign labour, roles, and responsibilities from the beginning. In other words, the initiative rapidly lost appeal, given the sense of uncertainty widespread among the participants in terms of what they were supposed to do. This was also due to the lack of regular meetings and a time-work schedule.

In this case, it is essential to contextualize the experience within a vulnerable and marginal condition that the inhabitants face daily. Although urban gardens were felt as a necessity for the committee, they may not have been perceived as such by many of the local inhabitants. During the focus groups, most of the residents had revealed to be keener on having playground areas for young kids, after-school common rooms, spaces for socialization, wi-fi areas, and other similar functions. All things considered, it is essential to re-prioritize the community's needs when promoting urban transformation and regeneration initiatives that are highly demanding for its inhabitants.

A drawback of the case study analysis is that, while it provides valuable insights into complex phenomena that would be difficult to study using other research methods, the selected projects are limited in their generalizability. The findings from one case study cannot necessarily be applied to other cases. Nevertheless, the replication logic behind the multiple-case study research, which must be distinguished from the sampling logic that requires statistical procedures, allows us to extend this research to further investigate other potentially relevant variables by increasing the number of case studies. A larger number of cases could provide a more comprehensive understanding of the phenomenon and a robust and more accurate basis for identifying patterns and variables that ensure the effectiveness and success of community gardens across different contexts.

Acknowledgments This work was performed under the grant "Città pubblica e nuovo welfare. La rigenerazione urbana per il diritto alla salute" (Protocol number: RM120172B3B15DB9) funded by Università La Sapienza di Roma.

References

Alvarez Mora A, Roch Peña F (2010) Integrated Urban regeneration in Europe summary document. Expert report. Instituto Universitario de Urbanística, Universidad de Valladolid. Instituto Universitario de Urbanística

Archdaily (2014a) "Esto no es un Solar": Reconvirtiendo parcelas vacías en espacio público [Parte I], https://www.archdaily.cl/cl/02-347483/esto-no-es-un-solar-reconvirtiendo-parcelas-vacias-en-espacio-publico-parte-i, last accessed 2023/11/06

Archdaily (2014b) "Esto no es un Solar": Reconvirtiendo parcelas vacías en espacio público [Parte II], https://www.archdaily.mx/mx/02-349303/esto-no-es-un-solar-reconvirtiendo-parcelas-vacias-en-espacio-publico-parte-ii, last accessed 2023/11/06

Archdaily (2016) Regeneración urbana: lo temporáneo como activador, https://www.archdaily.mx/mx/774436/regeneracion-urbana-lo-temporaneo-como-activador?ad_medium=gallery, last accessed 2023/11/06

Assemble Papers (2015) From concept to community at De Ceuvel in Amsterdam, by Katherine Sundermann, https://assemblepapers.com.au/2015/11/03/de-ceuvel-from-concept-to-community/, last accessed 2023/11/06

Atelier D'Architecture Autogeree, Agrocité-Colombes, https://www.urbantactics.org/projets/agrocite/, last accessed 2023/11/06

Barba Lata I, Duineveld M (2019) A harbour on land: De Ceuvel's topologies of creative reuse. Environ Plan A Econ Space 51(8):1758–1774

Battisti A, Calcagni L (2023) Avant gardening to grow the green city. In: Beyond the garden. theories, research, and projects for beautiful, sustainable, inclusive green urban spaces. Springer, Cham

Battisti A, Barnocchi A, Iorio S (2020) Urban regeneration for urban health. In: Urban health. AIMETA 2019. Green energy and technology. Springer, Cham, pp 43–56

Battisti A, Calcagni L, Calenzo A, Angelozzi A, Errigo M, Marceca M, Iorio S (2021) Urban health: assessment of indoor environment spillovers on health in a distressed urban area of Rome. Sustain For 13(10):5760

Benedict MA, McMahon ET (2012) Green infrastructure: linking landscapes and communities. Island Press, Washington

Calcagni L, Calenzo A (2023) Building a healthier living environment for people and the planet: a case study review. In: Mediterranean architecture and the green-digital transition: selected papers from the world renewable energy congress med green forum 2022. Springer, Cham, pp 87–101

De Ceuvel Homepage, https://deceuvel.nl/nl/, last accessed 2023/11/06

Environment & Society Portal (2019) StadtAcker, https://www.environmentandsociety.org/exhibitions/ecopolis-munchen-2019/stadtacker, last accessed 2023/11/06

Esto no es un solar homepage, https://estonoesunsolar.wordpress.com/, last accessed 2023/11/06

Ilieva RT, Cohen N, Israel M, Specht K, Fox-Kämper R, Fargue-Lelièvre A, Blythe C (2022) The socio-cultural benefits of urban agriculture: a review of the literature. Land 11(5):622

Le Parisien (2017) L'Agrocité déménagée de force de Colombes vers Gennevilliers, https://www.leparisien.fr/hauts-de-seine-92/colombes-92700/l-agrocite-demontee-de-force-a-colombes-20-02-2017-6695853.php, last accessed 2023/11/06

LMU (2020) The stadtAcker: Munich's most valuable Oasis, https://www.en.envstudies.carson-center.uni-muenchen.de/student-broadcasts/final-projects/2020/affected-by-climate-change1/index.html, last accessed 2023/11/06

Lohrberg F, Licka L, Scazzosi L, Timpe A, Verlag J (eds) (2016) Urban agriculture europe, vol 38. Jovis, Berlin

McGuire L, Morris SL, Pollard TM (2022) Community gardening and wellbeing: the understandings of organisers and their implications for gardening for health. Health Place 75:102773

Naumann S, Davis M, Kaphengst T, Pieterse M, Rayment M (2011) Design, implementation, and cost elements of green infrastructure projects. Final report. European Commission, Brussels, p 138

OECD (1998) Integrating distressed urban areas. Organisation for Economic Co-operation and Development

Quon S (1999) Planning for urban agriculture: a review of tools and strategies for urban planners. Cities feeding people series; rept. 28

R-Urban (2012) Projects. Agrocité, http://r-urban.net/blog/projects/agrocite/, last accessed 6 Nov 2023/11/06

U.S. Environmental Protection Agency (EPA) (2013) Questions about your community: indoor air. EPA New England, http://www.epa.gov/region1/communities/indoorair.html, last accessed 2022/07/05

Yin RK (2018) Case study research and applications. Sage

Chapter 10
Cultivating Cities: Urban Gardening's Impacts and Challenges

Silvia Cioli and Luca D'Eusebio

Abstract The humble, wise, and ancient act of planting tomatoes by those who obstinately take care of a void left over by urban sprawl is an ambitious and noble act contesting the multiple manifestations of injustice. It is a claim about the place people want to live in and the society people want to be part of. In recent years, urban gardening has become a practice which responds to many needs as it can promote a sense of community stimulating social cohesion; it strengthens urban ecosystems by increasing urban biodiversity and fighting climate change; it is a way of experimenting models of innovative production and management of public spaces, mitigating spatial and environmental injustice. The impacts are generally affecting positively on social, economic, environmental, political aspects of urban life, as for the challenges, there are still many since it is not easy to find the right balance between spontaneous and institutionalized initiatives.

Keywords Urban gardens · Community · Active citizens · Common good · Grassroot initiatives · Social cohesion · Inclusion · Gardening · Agro-biodiversity · Climate change · Public space · Innovation · Tradition · Environmental justice · Education

10.1 Cities, Public Space, and Agriculture

We live in times in need of a radical change especially in the urban context, where the world population is rapidly concentrating, that must face crucial issues. The issues cities must face are numerous and intertwined, they stretch from the environmental crisis, people's migration, pollution, public space restriction, just to mention a few. Added to this, the pandemic has modified many aspects of urban everyday life especially the use of public space. On one side, the pandemic has made us more dependent of internet, the virtual public space, where you rarely encounter anything

S. Cioli (✉) · L. D'Eusebio
Zappata Romana, Rome, Italy

A. Battisti et al. (eds.), *Urban-Rural Dialogue in Green City Design*, Designing Environments, https://doi.org/10.1007/978-3-031-88318-7_10

outside your "comfort zone"—in other words places where you meet what you like and what is like you. On the other side, the outdoor physical public space where critical fractures in society are manifest and where conflicts take place being it where you can bump into the diverse, the unpredictable, protests, strikes, etc. (Ratti 2022).

Therefore, when speaking about urban social ecological systems, we wish to focus on what we consider one of the urban major issues: the health of public spaces.

Public space enhances urban life and to have public, accessible spaces for all, is not something just 'nice to have', but it is a fundamental need for any global city and community to ensure social, cultural, environmental, and economic resilience. Its variety, quality, location, and accessibility are all important. In today's cities though, public space is drastically reducing. It is shrinking mainly because privatized for security reasons or meant for consumption. As an example, the city of Rome has reduced the access in areas within its central public park of Villa Borghese, in many *palazzi* courtyards and in the ancient forum (Ceen 2004). As Alan Ceen mentions:

> In Rome it is a rare fence to keep out cars. More often it is built to keep out pedestrians.

Urban community gardens are a response to the restriction and failure of contemporary public spaces.

Furthermore, we wish to point out a second issue regarding how agriculture seems to be increasingly coming back into the city. Numerous projects seek to enable cities in productive landscapes and cities grow interest in urban gardening. This reminds us how cities have historically been places where proximity, cultural diversity and mutual exchange have brought opportunities for social innovation and emancipation such as promoting urban gardening. However, the urban condition has also progressively become associated to unsustainable consumption practices, the greatest historical disconnection from the food system, and ongoing exploitation of humans and natural resources. This is locking both, urbanites, and farmers elsewhere, into a pervasive dependency from a socio-ecologically unjust food system (Tornaghi 2014) (Fig. 10.1).

The pandemic experience has showed us that the less cities rely on their rural hinterlands, the more they result unprotected to unpredictable shocks of their food supply chain. It is worth mentioning, as an example, how Lebanon has been promoting an 'Agricultural Jihad' (struggle), where politicians have invited citizens to plant and grow food anywhere, they can, for the strong economic crisis in the country added to the fact that about 80% of food is imported (Yee 2020).

Urban gardens seem to offer a synthesis between the traditional dichotomy urban/rural. They bring evidence on how agriculture has slowly started to make its way back to the city. After the separation of last century between food production and urban planning, the idea of urban life as antithetic to rural life was established. The food system seemed condemned to industrial and mass production. Urban gardens bring nature into the city in what are 'multifunctional' public spaces. This brings along innumerable benefits.

Urban gardens in the metropolitan fabric represent not only an exercise in the enhancement of public green spaces oriented toward nature conservation and

Fig. 10.1 Orti Urbani Monte Ciocci, Rome, credits: Zappata Romana

self-production of food, but these experiences also express an active and conscious citizenship that seeks to take ownership of the spaces and uses of its territory. Such experiences create a new form of participation and well-being in the realization of a common good through an ecological approach that includes the sustainable use of resources and crops and the building of relationships for the exchange of good practices. In this sense, the urban garden and the community that gravitates around it are a valuable testimony for the younger generations who can experience the sociality of growing together, exploring the relationship between nature and food in an environment that protects the importance of green space in the city and the need to respect it, learning to take care of something precious.

10.2 Active Citizenship

In recent years, urban gardening has become a practice which responds to many needs as it is can promote a sense of community stimulating social cohesion; it strengthens urban ecosystems by increasing urban biodiversity and fighting climate change; it is a way of experimenting models of innovative production and management of public spaces, mitigating spatial and environmental injustice (Certomà et al. 2019) (Fig. 10.2).

Fig. 10.2 Mandrione, Rome, credits: Zappata Romana

There is an ethical drive in 'doing' that is experienced in terms of enthusiasm and that enables one not to give up in front of difficulties and challenges that come along. In the simple 'doing' of those who take the responsibility of personally working stealing time from their private lives, many different actions are concentrated. These actions are hardly understood in their complexity. A community that takes care of its green spaces is imagined hoeing on a sunny day, period. Instead, that humble gesture of the cultivator or pruner or reclaimer of abandoned areas means much more: improving skills and sharing knowledge through informal channels; taking responsibility; putting oneself on the line with other people, betting to go down roads that are not beaten or safe, generally turning slightly 'against' as being ethically disobedient; cooperation, co-design with the ability to listen; ability to network. Little consideration is given to the fact that 'doing' has a lot of thought behind it.

A common trait of all experiences is to speak to the future, to place people at the centre of the revival, the relationships of communities with the territorial richness and diversity of environments, and an unprecedented, ambitious conception of culture, conceived as capable of generating a new and original project for the public space of communities. The manifestos with which these territorial cultural communities present themselves in larger arenas testify to their attempt to transform themselves from a dust to a system, from isolated laboratories to centres of cultural irradiation, capable of dialogue with the world. In the face of these ambitions, thinking only of remittance to the public sector is no longer sustainable.

In many cases, spontaneous urban gardens can be considered experiences of innovation. Being an innovator is the good side of the spontaneous action, while its downside lies in the uncertainty, the loneliness in being outside the institutionalized, the safe, the protected, the regulated, the codified. In fact, analysing Roman urban gardens it can be said that innovation seems to be linked to the spontaneity component of the phenomenon. To innovate, it is necessary to break out of the box. The challenge in most cities seems to be how to find a correct balance between spontaneity and institutionalized. Municipalities that promote urban gardening need to preserve the capability of a community to organize, plan and solve conflicts by itself as much as possible without interfering but at the same time, they should achieve being present and supportive towards the communities.

It is undoubtable that the protagonists are the citizens. To be precise, citizens not acting individually, rather gathered in communities in grassroot initiatives that utilize the garden activities to share resources, skills, and space. This is quite significant since it is the expression of a desire to act positively in neglected and abandoned urban areas as it happens in Rome. An optimistic action for the better. An improvement of one's neighbourhood for the sake of all. Most of all, a new consideration for public space that is not privatized but promoted and taken care of as a common good (Fig. 10.3).

10.3 Zappata Romana

Urban agriculture has always existed in Rome, still today there are remaining rural parts within the city. The city of Rome expanded in a disorderly manner leaving many unbuilt areas, some of which have become important urban parks preserving the well-known *Agro Romano* so appreciated by the *Grand Tour* travellers of the eighteenth century, other parts are instead being abandoned. A tradition for individual edible gardens, during World War 2 is also present till today, but the innovative phenomena of communities' urban gardens of the last decade represent an opportunity to do much more than just gardening.

Zappata Romana (ZR) starts in late 2010 when, during a participatory design process for a project in a suburban area of Rome, the involved inhabitants asked if they could take care of the new public gardens by themselves. As facilitators we wondered if this was possible in Rome and began an independent survey to verify the existence of bottom up and community led initiatives: we found 40 sites where citizens had spontaneously reclaimed abandoned areas utilized as new public space. We decided to make them visible on Google Maps, in a map based on icons with a brief description, an image and, when possible, a link for each site. This map helped the communities to know about each other. ZR is initially just the name of this map with the census of Roman shared gardens and vegetable gardens, spontaneous interventions characterized by the generosity of many citizens who by rolling up their sleeves had recovered abandoned areas to make them open public spaces for cultural, educational, and social activities. The map, accessible to all on Google Maps

Fig. 10.3 Guerrila Gardening at Corviale, Rome, credits: Zappata Romana

(more than 100,000 visits), and the work of ZR have helped the establishment of a collaborative network of the realities surveyed. ZR has contributed to the spread of this practice both with the website (where one can also find a manual on 'how to start', some videos of the protagonists and other materials) and through the organization of seminars and workshops, press campaigns and the massive use of social networks. In Rome, the shared gardens have grown from 40 in 2010, to 150 in 2013 (Fig. 10.4).

According to Stanley Matthews (Matthews 2012):

> Zappata Romana's map represents the city's unofficial gardens as a coherent urban landscape, as a whole, not as discrete and disconnected parts. The map endows these disparate gestures with an identity as an urban totality and outlines an alternative psychogeography of transgressive space in the city. Zappata Romana is helping to create an alternative mental map of the city, based not on monuments and transport routes, but on grass roots social interaction with the land. The ability to view the locations of all these sites simultaneously create a new urban consciousness, and psycho-geographic awareness of a whole new layer in the palimpsest, the lasagna that is Rome.

ZR's experience, developed over more than a decade, has practiced simultaneously the point of view of the observers of the many mapped experiences and that of those who rolled up their sleeves in 'doing' practice in an experience. The ZR association, which has been active in promoting the activation of abandoned areas, investigates shared gardens and vegetable gardens in Rome as a collective action of appropriation

Fig. 10.4 Coltivatorre, Rome, credits: Zappata Romana

of urban public space and the development of innovative environmental, cultural, economic, and social practices. These are areas recovered for collective use in difficult and mostly peripheral contexts, "places that don't matter", according to the definition of Andrés Rodrìguez-Pose, who defines them as the spaces abandoned by politics, those portions of the territory that have seen services rarefied to the point of reducing rights (Rodriguez-Pose 2017). From these places insecurity moves and stretches toward the centre. In contrast, the recovery of abandoned areas by citizens offers positive signs; a shift from a phase of nihilistic protest by atomized citizens to a new phase, in which the urgency to build an organized proposal for the renewal of social fabrics and spaces grows. The marginality of these places makes them unique spaces where community makers can experiment with new community relations, the only known antidote to degradation. Indeed, one of the effects of the growth of social and territorial inequalities is that the reduction of public proximity services, and the moving away from places of study and work are spreading into urban areas throughout Europe. It is no coincidence that the electoral program of Paris Mayor Hidalgo includes the proposal 'le Paris du ¼ d'heure', i.e., to make everything accessible from home in a quarter of an hour, services, amenities, jobs, for all city dwellers.

Historian John Foot (2019), in his latest book on Italy, and David Forgacs (2014) before him, suggests looking at the geographic and social margins of our country, the paranoid metropolitan peripheries, the groups of intersectional social disadvantage, weak families and foreigners, as the laboratories of contemporaneity.

ZR censuses, supports, and promotes active citizenship in the city of Rome by fostering the recovery of abandoned and degraded areas through the creation of urban gardens, in which the life projects of individual innovators, or the survival strategies of most, the missions of local associations and the functions of territorial government have amalgamated to become a shared project (Fig. 10.5).

The experiences mapped take the form of interventions for the material regeneration of abandoned spaces, such as disused agricultural areas, underutilized parks, areas intended for public services that were never realized, places on the margins of urban fabrics, so that they can become physical symbols of the aspiration of these collective subjects to transform themselves to something very similar to the 'parallel polis' of which Vaclav Benda (Benda et al. 1988) spoke, not a separate community that "lives in truth," but open to change and capable of accepting that the outcomes of this operation are not predefined, instead realized by learning and mutualistic communities and often real infrastructures capable of offering sociocultural services that can generate value and work. In this context, speaking about urban gardens, it is citizens united in collective subjects who take responsibility in those empty spaces left by politics and thought, as intermediary bodies between citizens and politics.

This 'empty' space is recognized and inhabited, filled with value by transforming it into a site, into the essence of architecture, which can be understood through Martin Heidegger's notions of building and dwelling, not as an end in itself, not as

Fig. 10.5 Orti urbani Garbatella, Rome, credits: Zappata Romana

mere shelter, but as a fundamental aspect of being human, which is to inhabit, to create a site on earth within which we can realize our existence spatially and temporally. Making a site gives presence and meaning to a place, and the ability to confer meaning is itself a source of social power and empowerment because they are places of shared social consciousness.

They are all united in these experiences by the idea that culture is a right of citizenship, and as such should have no barriers to access, not even economic ones. They show no nostalgia, and indeed sometimes marginality from the centres of the global economy is experienced as a space of freedom to be colonized, an opportunity to be seized.

10.4 Hortus Urbis

It was precisely by stepping outside the box that Hortus Urbis, an experimental project on public space and biodiversity run by ZR was born in 2012 in Rome's Via Appia Antica. The project was born with the activation of an unused green area in the frame of the Appia Antica naturalistic archaeological urban park by allocating it as a shared space and a garden with an educational vocation with plants used in ancient Roman times selected from those mentioned by Columella, Pliny the Elder, Cato, Virgil, and the discoveries of the excavations of Pompeii.

Since then, Hortus Urbis has been a living hub in the city where ideas and projects can be proposed and a relational environment to connect individuals, communities, and processes in collaborative ways. The project promotes synergies between different disciplines, such as: gardening, ecology, history, archaeology, botany, food, craft, and artistic productions (Fig. 10.6).

Hortus Urbis develops the dissemination of good practices, in addition to the historical research at ancient sources that continues as a work in progress, and aims to harmonize innovation, social impact and economic sustainability through: supporting informal groups, shared gardens and vegetable gardens, start-ups, experts, artists, illustrators, actors, experts, associations to define formats for workshops, courses and educational seminars based on the dissemination of good cultural and environmental practices; facilitating collaboration and encounter between different realities for the exchange of knowledge, contacts through sharing experiences. These entities together with families and children from neighbourhoods (and beyond), local associations, gardening newbies through the use of the garden for Sunday children's workshops, gardening, cooking, horticulture, painting and art courses and shared garden work, seize the opportunity of the open air public space constituted by the garden for the exchange of knowledge, experiences and weave new relationships seizing new job opportunities and determining an interesting opportunity for sociocultural innovation.

Hortus Urbis has given impetus to the construction of solidarity networks by involving public, private, and local community institutions for a common creative project that has seen the partnership of the Appia Antica Regional Park Authority

Fig. 10.6 Orti urbani Garbatella, Rome, credits: Zappata Romana

and other institutions, other institutional entities, foundations, foreign entities, associations, the Network of Shared Gardens and Gardens, specialized nurseries, professionals, and individual citizens.

The project began with an unusual initiative on the part of a public institution that involved citizens, through the association ZR, in the management of an albeit small portion of an area of strong historical and environmental value within a park.

The area hosts a shared vegetable garden intended as a common good of the community by applying a form of sharing economy suited to the realities and collaborative and participatory territorial networks present by implementing a different model of cultural management that based on meeting, leisure, exchange and sharing of knowledge with a high degree of resilience.

The synergies that are implemented increase individuals' possibilities for action and bring new relationships and collaborations. In fact, the garden is a ground for the exchange of knowledge and experience as a cultural product in real social events testifying that "it can be done." In this sense, Hortus Urbis produces "social innovation"because it transforms the experience of enjoying a park into a constant and engaging activity, linking the territory and the community to a concrete project, where people are directly involved in a path of care and maintenance of the natural environment.

The direct involvement of citizens for the recovery of the abandoned area to create the Hortus Urbis and its continuous care brings them closer to their own territory, to the park institution, putting within everyone's reach the opportunity for active citizenship with an immediate return for the collective well-being by improving everyone's living environment.

The shared physical space and its care, the activities that take place there, characterized by active participation, constitute a presidium and enrichment for the social capital of the territory for the opportunities to weave direct social relations by coagulating everyone's experience around common experiences and values and fostering intergenerational relations, social inclusion and the exchange of experiences and knowledge by increasing everyone's possibility of action. It is a project that makes it possible to broaden and extend the ability to involve the different social components of the area, rebuilding cultural identity and making it a collective heritage.

10.5 Urban Garden's Impacts

The impacts of the more than 200 Roman urban gardens as reported on the ZR map are like many other cities. The peculiarity of Rome is the active citizenship drive with the presence of many associations as a reaction to the problematic and unsuccessful maintenance of green areas and weak collaboration of the local authorities. Nevertheless, the heterogeneity in purposes, shapes, and management of urban gardens initiatives, have a lot in common among the different cities and countries. The common impacts can be summarized in four main thematic areas (Eizenberg 2013) regarding: the human experience; politics; urban economy; the physical environment.

The human experience, in terms of personal and communal impacts fosters a growth of sense of community and social cohesion. More than a place for growing things, the urban garden can foster collective labour and informal social interactions that lead to a sense of belonging, connection to space, solidarity, and an improved security in a neighbourhood. Furthermore, its well-known educational potential offers non-traditional and attractive hands-on approaches for teaching outside the classroom. An urban garden is an opportunity to share resources, skills, and space for different issues beyond environment, ecology, nutrition, and science, for all ages. The contribution to phycological wellbeing and health when spending time outdoors, actively immersed in nature, or even passively, is beneficial as demonstrated by many scientific studies. An active experience in nature, as gardening, is even more rewarding, added to the outdoor physical exercise and improved nutrition. It also produces opportunities for leisure and socialization.

Gardeners learn that neighbourhood-oriented actions have political implications that can help change the city, improve public space, and promote food justice. The involvement of marginalized groups and transformation of urban spaces against privatization of space can be a resource for the mitigation of spatial and environmental injustice. To take part in an urban garden is a manner to resist to the constant restraint of public space in most cities either due to speculative logics or security reasons. Communities can promote a great variety of uses in the urban sites they are responsible for, introducing innovative models for the production and the management of public space. To grow food for self-sufficiency and sustenance can be the main issue typically for cities in Southern countries but also for cities of the Western world facing crisis periods, as recently proven during the covid pandemic, contributing to enhance food availability, accessibility and affordability (Fig. 10.7).

The act of cultivating in an urban context affects the urban economy too. *Grow-it-yourself* also means finding one's own expression in the products of one's labour. It means setting oneself apart from a life of consuming objects of industrial production. Seeking individual expression is also a quest for new forms and places of community. Furthermore, sharing economy in community gardens heightens community engagement and brings a sense of social connectedness. To be a part of the sharing economy goes well beyond the rational arguments of saving resources and money. The experience of meeting, working, and sharing with others in a community brings joy and excitement. By encouraging people to participate, urban gardens gather and combine a large amount of knowledge in productive ways. Since there are usually no agricultural professionals among the gardeners, everyone depends on whatever knowledge is available and everyone is open to learning. We cannot have a common resource without a common strategy for managing it (Ostrom 1990), the commons require a set of rules and seem to offer an alternative to the battle between public and private. Knowledge commons too since open source is the central guiding principle in all community gardens. As mentioned, common goods are an expression of

Fig. 10.7 Villaggio 95, Rome, credits: Zappata Romana

the changing relationship between the public and the private. Open source is the central guiding principle in all community gardens where people are encouraged to participate. Urban gardens gather and combine a large amount of knowledge in productive ways. Everyone takes advantage of whatever knowledge is available and everyone is open to learning. Another aspect concerns the vacant urban sites that are transformed, when reclaimed by the community's efforts, into highly used and productive spaces. This can also impact on the real estate value of urban areas near the urban gardens increasing the value of their surroundings.

To cultivate and maintain reclaimed abandoned sites is also a way to increase urban agrobiodiversity and fight climate change. Urban gardens are microenvironments that contain high levels of species diversity with a rich variety of wild plants and traditional local edible vegetables reflecting the cultural identities of the gardeners. They can address climate change issues such as rising temperatures, increasing rainfall, flooding, and urban food insecurity. In other words, the physical environment can be positively affected as the urban ecosystems are being strengthened, but also for the regeneration strategy of many unbuilt and uncertain areas of the city. Urban garden can be an opportunity against urban sprawl where vacant urban sites are transformed into highly used and productive spaces and an opportunity to regenerate urban soil through soil improvement and green usage of organic solid waste by using composting methods (Fig. 10.8).

10.6 Urban Garden's Challenges

It is crucial that city authorities and planners start recognizing the importance of urban gardening, frequently only considered as mere maintenance of underutilized land, by supporting it as a strategy to be used not only as reaction in times of crises but mostly as a precious tool to be part of a global strategy that can enhance the resilience and sustainability of urban areas and inhabitants with communities that want to contribute to addressing climate change. In many cities as it is happening in Rome, the phenomena are growing, the waiting lists of each garden get longer, and more people wish to change stile of life. Between the legalities and the challenges of soil and weather, urban agriculture does face difficulties. Local authorities should be able to plan, site, negotiate, set rules, and protect gardens and more specifically: figure out what types of systemic planning tools are available to integrate planners, practitioners, and civil society in a process of joint learning that can guide the development of urban agriculture more effectively; design an urban garden program as an integral part of a city's park system or redevelopment projects through a wise combination of municipal needs and visions with the interests and needs of a community; valorise the vital mix of activities like public events, exchanges, opportunities, learnings that can take place; consider that there cannot be not only volunteer engagement, there is also a need for professional coordination; not undermine aesthetic issues; understand the needs for specific local contexts such as the resources, the possible partners, suppliers, which roles can be performed; incentives for soil

Fig. 10.8 Hortus Urbis, Rome, credits: Zappata Romana

testing services for gardeners, the development of real-time soil quality map of the city or the study of potential pollutant sources near the gardens; work in coordination with local public actors such as schools, universities, research centres, prisons, hospitals, public libraries, etc. and also local private actors like farmers, markets, foundations, etc.; build communication tools to reach as many citizens as possible, coherent with a global strategy and vision for the city's sustainable development, food policies and governance; build a metropolitan food strategy to reduce health inequalities, negative environmental impacts of the current food system, support a vibrant food economy, promote a city's food culture, improve overall food quality and develop food security.

Going back to the starting point, reflecting about what we learned from the pandemic together with the considerations about the outdoor physical public space where critical fractures in society are manifest is where conflicts take place being where you can bump into the diverse, the unpredictable.

In the city each one seems to have a separate 'place' where to be, as schools, offices, theatres, etc. Even in parks, where everybody goes, the elderly are probably sitting in benches, kids are in a playground, some adults are jogging, in other words there is no interaction.

But there is a space, in the physical realm, within the city, where people interact, share, where differences converge and where barriers are overcome … this unique urban public space is: the urban garden.

References

Benda V, Simecka M, Jirous I, Dienstibier J, Havel V, Hejdanek L, Simsa J, Wilson P (1988) Parallel polis, or an independent society in Central and Eastern Europe: an inquiry. Soc Res 55(1–2):237–243. https://doi.org/10.1007/978-1-349-11117-6_2

Ceen A (2004) Roma Cancellata. Public areas inaccessible to the public http://www.studiumurbis.org/assets/catalogs/Roma_Cancellata.pdf

Certomà C, Sonderman M, Noori S (2019) Urban gardening and the struggles of social and spatial justice. Manchester University Press, Manchester

Eizenberg E (2013) From the ground up: community gardens in New York City and the politics of spatial transformation. J Polit Ecol 21(1):555. https://doi.org/10.2458/v21i1.21153

Foot J (2019) L'Italia e le sue storie 1945–2019 (The Archipelago: Italy Since 1945, 2018), traduzione di Enrico Basaglia, Collana I Robinson. Letture, Laterza Roma-Bari

Forgacs D (2014) Margini d'Italia. L'esclusione sociale dall'Unità ad oggi, Laterza, Bari-Roma

Matthews S (2012) "Zappata Romana" and spontaneous urbanism in Rome, Northeast modern languages conference Rochester, New York

Ostrom E (1990) Governing the commons: the evolution of institutions for collective action. Cambridge University Press, Cambridge. https://doi.org/10.1017/CBO9780511807763

Ratti C (2022) Il futuro delle città: lo spazio pubblico torna protagonista della società', la Repubblica 22/07/2022

Rodríguez-Pose A (2017) The revenge of the places that don't matter (and what to do about it). Camb J Reg Econ Soc 11(1):189–209. https://doi.org/10.1093/cjres/rsx024

Tornaghi C (2014) Critical geography of urban agriculture. Prog Hum Geogr 38(4):551–567. https://doi.org/10.1177/0309132513512542

Yee V (2020) 'Agricultural Jihad': a hungry Lebanon returns to family farms to feed itself', New York Times, 05/09/2020, New York

Zappata Romana https://www.zappataromana.net

Chapter 11
Constructed Wetlands as Green Infrastructure to Improve Quality of Life

Ahmed Abouaiana

Abstract Green Infrastructure (GI) is vital in enhancing the built environments' ecosystems. It provides numerous benefits (environmental, social, climate change mitigation and adaptation, and biodiversity); GI has different artificial features, such as building green roofs, eco-bridges, or constructed wetlands (CWs). CWs are low-cost, nature-based solutions to treat wastewater and improve water quality. Beyond the technical solutions, this chapter highlights the role of multifunction CWs in improving ecosystems and quality of life in urban areas. To pursue this, the chapter investigates case studies; all are pilot projects in their contexts (Egypt, Italy, France, and Denmark) and have different CW characteristics and scales: limited architectural and landscaping practices integrated with spatial planning and city development. Furthermore, the approaches and stakeholders are described. The results demonstrated that CWs and GI are practical and effective tools that can maintain biodiversity, contribute to sustainable development, address climate change, offer pleasant environments, and support quality of life. The role of architects and urban planners is highlighted. More interdisciplinary approaches and innovative GI solutions that focus on citizen's well-being are still required.

Keywords Sustainability · Nature-based solutions · Ecosystem-based solutions · Architecture and urban planning · Health and well-being

11.1 Introduction

Climate change threatens the world with rising temperatures, rainfall pattern changes, and extreme weather events. Urban areas are most vulnerable due to population, infrastructure, and property. The developing world is most affected, and the global north emits the uppermost greenhouse gas emissions. The Intergovernmental Panel on Climate Change reports confirm the accelerating impacts, and the current

A. Abouaiana (✉)
Department of Architecture, Faculty of Engineering, Sinai University, Ismailia, Egypt

A. Battisti et al. (eds.), *Urban-Rural Dialogue in Green City Design*, Designing Environments, https://doi.org/10.1007/978-3-031-88318-7_11

action is insufficient. Additionally, the global urban population is estimated to expand by 2.5 billion persons between 2018 and 2050 (UN 2019), rapid global urbanization has adverse impacts on global environments, natural resources, and human well-being (Navarro et al. 2020; Wang et al. 2020), and stress the infrastructures, urbanization goes hand in hand with infrastructure (Corpuz 2018). Thus, urbanization has numerous pros that can promote economic growth if it is well-managed (Dobbs et al. 2012), especially with its position with infrastructures. In this light, the global community pays exceptional attention to green infrastructure (GI). The European Union in 1992 founded the network of nature protection areas, so-called Natura 2000, as a centre of European GI. Briefly, GI is defined as "*the network of natural and semi-natural areas that are managed to deliver ecosystem service, including water purification, air quality improvement, quality of life, well-being, and climate mitigation.*" (EEA 2021; EU 2021).

GI is the most cost-effective and better and cheaper option for spatial planning, a fundamental tool for the adaptation of cities in order to achieve resilience and sustainability, quality, health, and environmental safety objectives because it enables the achievement of a range of ecosystem services that improve ecology, society, and the economy, besides supporting tourism, decrease insurance premiums for businesses and dwellings, enhance landscape maintenance, natural beauty, and protect biological systems (Benedict and McMahon 2012; Gulati and Scholtz 2020).

In the same context, the EU Green Deal prioritizes climate-proofing, resilience-building, prevention, and preparedness via nature-based solutions (NbS) to achieve primary EU policy objectives: to foster biodiversity and enhance climate resilience. The European Commission glossary defined the NbS as "*Cost-effective solutions that achieve environmental and socio-economic benefits help build resilience and integrate nature into urban and natural areas through systemic interventions*."

Hence, greening and naturalizing existing/new infrastructures instead of traditional infrastructure (grey) and promoting NbS (the issue is gaining momentum) can address climate change and environmental challenges; examples of these solutions are eco bridges for wildlife crossings, urban tree canopies for cooling and shading, preamble paving and rain gardens to reduce stormwater runoff, and green walls and roofs to improve microclimate and biodiversity (Fig. 11.1) besides

Fig. 11.1 Some green infrastructure elements (Source: Unknown authors, licensed under CC BY-NC, Microsoft Online Pictures)

constructed wetlands (European Commission, Directorate-General for Environment 2014; Bertule et al. 2014).

11.2 What Are Constructed Wetlands?

Wetlands are such an ecosystem saturated with water; they are the zones where the water predominates flora and fauna; that happens when shallow water covers the land or the water surface becomes near to the ground, less profound than six meters (Ramsar Convention Secretariat 2013). Likewise, the CWs are low-cost (engineered) that simulate the mimic procedures found in the natural ones to improve water quality and resource replenishment by increasing the number of surface water supplies, recharging groundwater, and removing pollutants from urban runoff (Bai et al. 2020). The subject has been debated in Europe for three decades (Haberl et al. 1995). UN-Habitat (2008) provided a detailed technical manual on constructing, installing, and functioning CWs).

Briefly, CWs promote several benefits; they can filter pollutants and excess nutrients to improve water quality in ecosystems, offer recreational opportunities and natural settings for outdoor activities (e.g., bicycling, photography, and bird watching) that benefit human well-being, and support biodiversity for an expansive variety of fauna and flora, and ecotourism. Meanwhile, they promote educational activities and serve as outdoor classrooms where people can learn about their importance (e.g., sustainable living practices) (DNDA n.d.), also promote aesthetic values and enhance the beauty of a community, providing a peaceful and natural space for residents to enjoy more pleasant living urban environments (Do and Kim 2020). Figure 11.2 illustrates the Tiber Island urban wetland in Rome.

11.3 Multifunction Land Use CWs: Architecture and Urban Planning Perspective

The nature of ecosystem-based solutions like CWs is predominated by environmental engineering and water management; however, architecture and urban planning domain scholars, in recent years, have generated a tendency to multifunctioning these solutions to function as urban parks and public spaces as an accelerator to sustainability, climate change mitigation, and improve citizens quality of life (QoL), as the world felt the role of urban parks, especially during and after COVID-19 as discussed significantly recent 3 years.

Multifunctionality refers to integrating different land uses on a single piece of land or implementing a single multifunctional land-use type; the various goals must be coordinated and achieved simultaneously or successively (Zhang and Schwärzel 2017) to multiply functions in the same area. Likewise, multifunctioning CWs

Fig. 11.2 Tiber Island urban wetland strongholds biodiversity and ecotourism. Photography: Ahmed Abouaiana

within urban planning and architectural practices; many studies promoted CWs as urban parks and landscape applications in developing and developed countries to tackle socio-economic and ecological factors and achieve urban resilience, and mitigate climate change (Zhu and Wang 2022; Gaber 2020; Haron and Feisal 2020). However, a recent study by Lachinani Kordesofla and Abna (2023) confirmed the lack of studies to consider CWs as landscape functions and to be integrated into planning and that considered aesthetic aspects, while they descriptively discussed and overviewed 25 CWs under different climatic circumstances worldwide.

In order to foster a new approach, this chapter highlights the essential need to integrate NbS and GI, like CWs, into architectural and planning practices to provide a trajectory for the architects and planners, supporting their capabilities to integrate ecosystem-based solutions into design processes, at different scales (from building to urban spaces) to promote citizens' QoL and enhance sustainability, besides emphasizing the primary function of these eco-based systems. Simultaneously, it highlights the stakeholders' responsibilities and roles regarding the field's nature.

To pursue that, the chapter uses the descriptive method that explains the reality of CWs as a GI element of the urban space by theoretically viewing their function, benefits, and associated challenges. Then, it presents pilot case studies with various characteristics developed from top-down and bottom-up perspectives in different

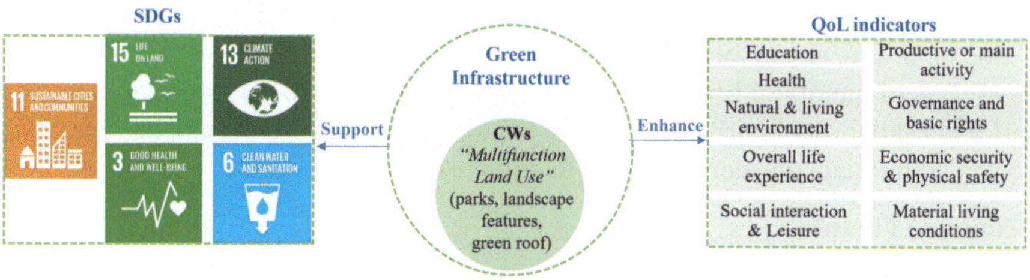

Fig. 11.3 The interaction of multifunction CWs—GI, as architectural, landscape, and urban planning tools to support QoL and SDGs

contexts and scales in Egypt, Italy, France, and Denmark. Finally, it shows a running project under development by the author, employing micro-scale GI.

11.4 CWs Benefits and Challenges at a Glance

The CWs contribute to the overall QoL, which the World Health Organization defines as "individuals' perception of their position in life in the context of culture and value systems in which they live and concerning their goals, expectations, and standards." (WHO 2012). European Commission identified aspects of QoL that can be measured statistically (8+1 dimensions) related to functional capabilities, self-defined well-being, and personal life satisfaction.

Inversely, CWs can support sustainable development goals (SDGs), namely SDG3, Health and Well-being (e.g., promoting physical activity, mental relaxation, and social interaction). SDG 6, Clean Water and Sanitation (e.g., improve water and sanitation management, protecting water-related ecosystems). Goal 11, Sustainable Cities and Communities (e.g., provide inclusive, safe, resilient, sustainable, and climate change-adaptable urban environments). SDG 13, Climate Action (mitigate climate change, absorbing carbon dioxide, and improving urban microclimate). SDG 15, Life on Land (e.g., protect ecosystems, preserve natural habitats, and provide essential services like air purification and water filtration). Figure 11.3 illustrates the juxtaposition of multifunction CWs as GI with QoL and SGDs.

Although there are vast benefits to implementing CWs, the associated challenges of implementing it and GI can be summarized based on references. (Makopondo et al. 2020; U.S. Environmental Protection Agency 2024; Velasco et al. 2023), as follows:

Technical Aspects

- Effective design is decisive for the accomplishment of CWs for treating wastewater and wildlife habitat provision.
- Frequent maintenance is essential to preserve CWs functioning properly.

- Considering their specificities area depending on accessibility, for example, the location and extent of wetland integration in residential areas determine the aspects of quality of life most satisfied (Pedersen et al. 2019).
- Land availability in urban or densely populated areas poses a challenge for CWs and additional considerations due to climate conditions.

Economic Factors

- High initial and running (maintenance) costs.

Regulatory Approval

- Acquiring the required approvals from regulatory bodies can be a challenge subject to environmental regulations, restrictions, and spatial planning.

Social Aspects

- Related to stakeholders, what is the acceptance level of resistance from the public or local stakeholders unfamiliar with the concept of CWs or have concerns about potential odours or other perceived negative impacts?
- Architects' and urban planners' awareness of the CWs' role (and GI generally) as a landscape element can be employed at different scales.

The latter factor is considered the most crucial factor, beyond the technicalities of CWs, which is predominated by environmental engineering (water management). Thus, there is a crucial need to determine the level of community engagement in different life cycle stages of implementing CWs (early planning, implementation, and maintenance). On the other hand, engaging architects, landscapers, and urban planners will ensure the achievement of the CWs' objectives.

Therefore, participatory approaches are necessary for effective urban green infrastructure planning and the design, implementation, and assessment of NbS, like CWs. Engaging with different stakeholders to transfer knowledge between actors while simultaneously addressing potential barriers is essential to gaining societal acceptance. From the top-down perspective, local and regional authorities play a vital role in this process. They should engage in strong horizontal and vertical cooperation, linking with the private sector (logistics and technical support) to ensure successful implementation (Climate ADAPT-EEA 2023).

Additionally, encourage local groups (residents, park users) to become wetland stewards to foster a sense of ownership (Zeng and Deal 2023) by engaging them before developing proposals in the design process, maybe via interviews and questionnaires to inform design decisions regarding recreational, social activities and awareness programs to enhance relevant educational activities (Vargas-Hernández and Zdunek-Wielgołaska 2020). The NGOs can facilitate this and sustain the maintenance and monitoring of the project. Other stakeholders that will need to be consulted include, for instance, local water companies and environmental agencies (e.g., wildlife) (ENFIELD Council and ZSL 2021).

The author advocates engaging the architects and urban planners within the core expert team and vice versa; from a bottom-up perspective, they are encouraged to embed GI within their architectural design practices, like green roofs and walls, to

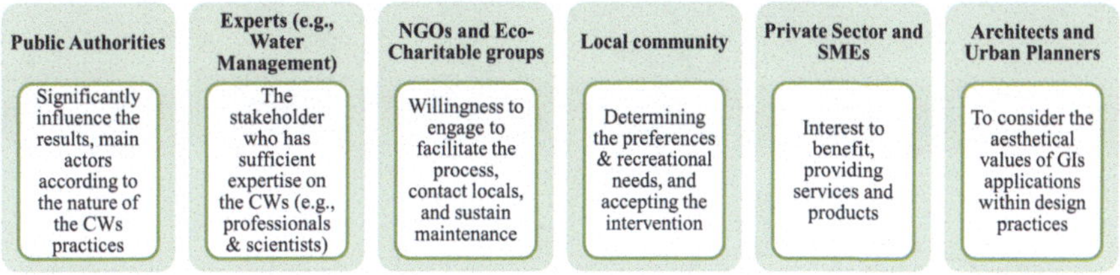

Fig. 11.4 Stakeholder profiling, based on the GI domain characteristics, with advocation to engage architects and urban planners as main actors in the process

support the ecosystem and improve building's environmental performance, in addition to integrated it within the landscape in different projects, as well as urban planning practices, showing the benefits to the project owners and investors; this will not occur without the awareness levels that this chapter tries to support. Figure 11.4 shows profiles of the stakeholders engaged in multifunction CWs, highlighting the role of architects and urban planners.

11.5 Case Studies

11.5.1 Mixed Approach: Tenth Ramadan Wet Land Park

The first one is a pilot-constructed wetland project (adjacent to a water treatment plant) in Tenth Ramadan City, in the Suez Canal region (eastern of Cairo), the approximate area is 0.037 km^2 (1050 m × 35 m), as the first constructed wetland urban park in Egypt. The aim was to combine landscape features and ecological functions (water treatment) by establishing a "happy public park." The former land use was vacant land. Figure 11.5 shows a recent top view on Google Maps. The approach was mixed top-down and bottom-up (initiated by architects); the project is interdisciplinary and developed by a collaboration of stakeholders consisting of project partners, namely:

- Cairo Higher Institute initiator (architecture and planning).
- Professionals' collaborators:(architects-leaders, water treatment, landscape designers, and agriculture, electro-mechanical, and civil engineers.
- Scientists (university professors) in workshops.
- Technical Support: National Water Research Centre.
- Public bodies: the Tenth Ramadan City Council representatives, New Urban Communities Authority.
- The Science and Technology Development Fund (STDF), for funding.

Recent studies by the lead researchers to quantify the environmental, social, and economic aspects based on discussions among professional stakeholders (interviews and workshops); simultaneously, they provided a guideline for establishing

Fig. 11.5 Google Maps show the project's progress (Source: Imagery ©2023, Airbus, CNEs, Maxar Technologies, Scankort, Map data ©2023)

Fig. 11.6 The sketches developed by the workgroup based on conceptual design provided by the research team committee

such local projects, and they examined the locals' visual and perceptual preferences for landscape elements (Barmelgy et al. 2022; Omayer and Haron 2022).

Notwithstanding that, the author has contributed to a two-day workshop in the early planning stage of this project as an expert architect and academician within one of the workshop teams consisting of landscapers, architects, and urban planners to develop the outcomes. Figure 11.6 shows initial sketches while brainstorming among the team members, while Fig. 11.7 depicts the key considerations developed by the team during the workshop. The author promoted some ideas inspired by the management experience of Caffarella Park in Rome.

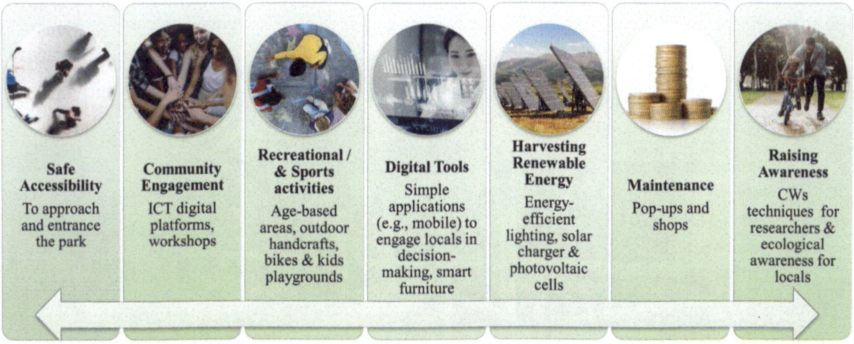

Fig. 11.7 Key considerations by the author's teamwork within the workshop. Developed by the author based on Abouaiana, 2021

11.5.2 Top-Down Approach: Gorla Maggiore Park, Italy

The second project is Gorla Maggiore Water Park, in Gorla Maggiore (suburban, according to Degree of Urbanization classification by Eurostat) Milan, the Lombardy Region, Italy, and it is adjacent to the Olona River. The total area is 30,000 km^2; the former land use is for poplar plantations. The project implementation started in 2008 and was inaugurated in 2013 (as the first in Italy in this form of GI). The main aim was to "test the feasibility of a green infrastructure, instead of a traditional grey infrastructure." (Oppla 2023); the water park covers include a set of CWs to improve water quality, protect against flooding, and protect biodiversity, besides enhancing social interaction, promoting green recreational areas (comprising 70% of the total area), and educational activities. The project responded to the EU Water Framework Directive, funded by the EU Project, EU FP7 project OpenNESS.

The approach was top-down, the project is interdisciplinary, and the collaborators and partners have been described by Liquete et al. (2016). Stakeholders were selected to cover all those affected by the project and those who can implement it (their levels of interest and influence), including crucial water managers and decision-makers, experts, and representatives of social groups and potential area users, and they are:

- Lombardy Regional Authority and Gorla Maggiore municipality.
- Expert in a cross-disciplinary engineering firm (IRIDRA) consisting of with a particular experience in sustainable water management.
- Scientists (university professors).
- Calimali Association NGO and Cariplo Foundation (Charitable) to connect with locals, facilitate implementation, provide maintenance, and promote activities.

One of this project's findings is that the locals are "willing to pay" triple for GI than for the conventional, and much more if it is bounded by a recreational park, emphasizing the essential need to expand the application of these projects. (Liquete et al. 2016; Masi et al. 2017; Rizzo et al. 2021) (Figs. 11.8, 11.9, 11.10 and 11.11).

Fig. 11.8 Playground made from local materials to improve physical and mental health. Photography: Ahmed Abouaiana

Fig. 11.9 Sport activities, e.g., cycling to improve physical and mental health. Photography: Ahmed Abouaiana

Fig. 11.10 Wetland to support biodiversity, recreational activities, and aesthetic values. Photography: Ahmed Abouaiana

11.5.3 Top-Down Approaches at Mega Scales in Paris and Copenhagen

In France, the Ministry of Ecological Transition promoted the eco-district approach (ÉcoQuartier), supported by promoting new sustainable ways to create and operate the city; in other words, "it is a development project that integrates the issues and principles of the city and sustainable territories; meanwhile, it supports stakeholders from all territories to design, build and manage cities and territories that meet the SDGs, via integrating all the issues of sustainable development and offers its residents the possibility of a lifestyle that meets the significant challenges of climate, biodiversity, and living together, relying on local resources." (Ecoquartiers 2024).

The City of Paris launched a pilot eco-district project in 2002, Clichy-Batignolles district, with the aim to provide eco-districts morphed into Paris first build as a model of sustainable development for the rest of the city. The former land use was an industrial wasteland. Among the sustainability strategies, it employed green spaces on rooftops, outdoor spaces, and parks to manage and collect rainwater, then channel it towards constructed wetlands (40% of the park's watering needs are met by rainwater), improving microclimate, supporting biodiversity and ecosystems, and generating recreational realms. Figures 11.12 and 11.13 show the contribution of one of the CWs within the district to support ecosystems and citizens' QoL.

Fig. 11.11 Information signage provided by the NGO highlights the different roles of Gorla Maggiore Water Park. Photography: Ahmed Abouaiana

Fig. 11.12 GI as a tool to manage runoff and rainwater and provide recreational activities for different age groups, such as scootering. Photography: Ahmed Abouaiana

Fig. 11.13 Constructed Wetland supports biodiversity and ecosystem and offers recreational and physical activities (photography, duck watching, and fitness). Photography: Ahmed Abouaiana

In Denmark, the City of Copenhagen, in response to the Copenhagen Climate Adaptation Plan in 2011 to deal with the environmental challenges, designed the Orestad district (within the urban development national plan) to be built with water by multifunctioning the CWs for stormwater management linked the aesthetic purposes as landscape features and support citizen's' QoL. The design concept is to clean runoff from rooftops diverted directly to the canals, while the less clean runoff from roads is collected in a separate system and treated with green technology to clean the water locally before entering the aquatic environment, besides reusing rainwater from the canals to irrigate parks and road trees. This shows how the CWs are an integrated element within the holistic design process for the new district to achieve urban resilience and address environmental challenges (formerly, the same pipes used for treating rainwater, road water, and wastewater), (Orestad 2024), that causing devastating impacts. Figures 11.14 and 11.15 depict integrated CWs in spatial planning of the residential and commercial areas in the Orestad district and their contribution to aesthetical aspects.

Fig. 11.14 Outdoor sitting areas in a residential zone along the constructed wetland in Orestad district, Copenhagen. Photography: Ahmed Abouaiana

Fig. 11.15 Constructed wetland within a commercial area in Orestad district, Copenhagen. Photography: Ahmed Abouaiana

11.5.4 Bottom-Up Approach: Downscaling into Architectural Practices, Egypt

The author developed an architectural project as a service yard (front offices) for EDARA Property Management firm; the project is in Sheikh Zayed City, Greater Cairo Region. The project's main aim is to accommodate business employees (business needs), with the objectives of providing a pilot suitable prototype of the firm and supporting the welfare of employees. The project employed several active and passive design strategies, including micro-GI elements, namely, CWs to treat toilet wastewater and reuse greywater into irrigation; the CWs system was developed based on a similar implementation to treat wastewater of a building in the University of Bologna, Italy (Lavrnić et al. 2020), the stakeholders were EDARA firm who owns the project, meanwhile to provide technical experts from the departments of maintenance and landscape (electrical engineer, plumbing engineer, irrigation engineer, and landscape), in addition to the author who is the lead architect designer of the project, who promoted the idea and discussed the technicalities with EDARA's experts. In a nutshell, the simulation (provided by the expert) showed that the potential daily drainage capacity of greywater is 9.7 m^3 (the total is 14.2 m^3); the greywater to be used in irrigating green areas and injected into an artificial lake (as storage and to improve outdoor microclimate). The team provided additional studies (civil work, supported pumps, and solar energy generation). The project is ongoing (design phase), and the estimated implementation period in quartile 3 of 2024 (Fig. 11.16).

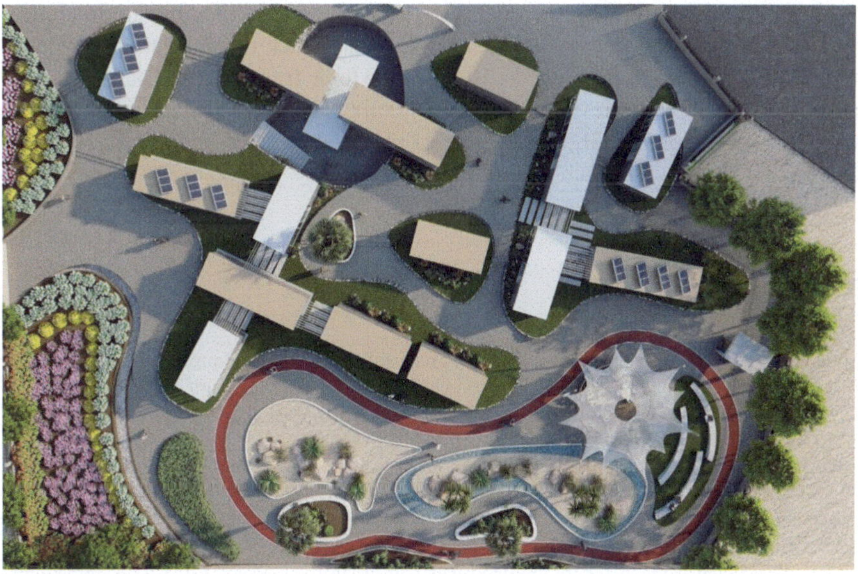

Fig. 11.16 The project's design concept shows the distribution of the artificial wetland as a micro-GI integrated with sustainable strategies designed by the author

11.6 Conclusion

Urbanization is a global phenomenon that has expanded space and altered the natural landscape. Building resilient communities requires providing innovative solutions to altering the conventional infrastructure with GI and NbS and maximizing the benefits by multifunctioning land uses for multi-objective optimization, not only addressing environmental challenges but also providing vast ecological, cultural, and economic values, flourishes flora and fauna, and improving citizens' well-being and QoL. Implementing pilot NbS projects is distinctly inevitable to protect communities and fill knowledge gaps (Castelo et al. 2023).

The nature of the domain is characterized by top-down approaches and complexity, represented by engaged stakeholders and interlocking requirements (e.g., land availability, cost, and technical capacities), showing the role of central decision-makers in facilitating establishing GI. On the other hand, the chapter advocated that creating momentum from professional architects and urban planners to downscale GI applications into architectural practices is essential, so the architects, as the author is, are encouraged to widen their expertise area to promote micro-scale and low-cost NbS, like, applying green walls and roofs as to improve ecosystem, well-being and improving building's thermal performance. NbS is being used for wastewater treatment (alternative to CWs) at the building scale (Addo-Bankas et al. 2021). This might answer the question: who should initiate GI? And how?

Creating liveable environments with a focus on quality of life remains an essential priority to social acceptance of the NbS, so relationships between developers, architects, and scientists, local authorities who can work together to take joint responsibility for improving the quality of urban life and achieving sustainability, should be fostered beyond the technical solutions, which the chapter tried to promote. Finally, although the chapter discussed the topic from a limited perspective, it can be integrated with previous literature. Future research is still required to simplify the applications of GI, and the role of public authorities is needed to support GI initiatives at downscales.

References

Abouaiana A (2021) Constructed wetlands as green infrastructure to improve quality of life. In Conference: the 6th architecture & environment international symposium entitled cities, green systems and human health and well-being. Rome on 9–10 April 2021, Sapienza University of Rome. https://doi.org/10.13140/RG.2.2.25794.50888

Addo-Bankas O, Zhao Y, Vymazal J, Yuan Y, Fu J, Wei T (2021) Green walls: a form of constructed wetland in green buildings. Ecol Eng 169:106321. https://doi.org/10.1016/j.ecoleng.2021.106321

Bai X, Zhu X, Jiang H, Wang Z, He C, Sheng L, Zhuang J (2020) Purification effect of sequential constructed wetland for the polluted water in urban river. Water 12(4):1054. https://doi.org/10.3390/w12041054

Barmelgy MEL, Elsaman SM, Refaat DT, Haron A, El Maalt AE (2022) Constructed wetland park as happy public space to achieve quality of life: case study of 10 Ramadan city. IOP Conf Series Earth Environ Sci 992(1):012009. https://doi.org/10.1088/1755-1315/992/1/012009

Benedict MA, McMahon ET (2012) Green infrastructure: linking landscapes and communities. Island Press. https://books.google.com.eg/books?hl=en&lr=&id=2xTJvYqzFNkC&oi=fnd&pg=PR5&dq=green+infrastructure&ots=3pSeeKSNQx&sig=tLKLtqPGZB3-X84gAYjNbRslh1Y&redir_esc=y#v=onepage&q=greeninfrastructure&f=false. Accessed 29 June 2024

Bertule M, Lloyd GJ, Korsgaard L, Dalton J, Welling R, Barchiesi S, Cole R (2014) Green infrastructure guide for water management: ecosystem-based management approaches for water-related infrastructure projects. United Nations Environment Programme, Nairobi

Castelo S, Amado M, Ferreira F (2023) Challenges and opportunities in the use of nature-based solutions for urban adaptation. Sustain For 15(9):7243. https://doi.org/10.3390/su15097243

Climate ADAPT-EEA (2023) Urban green infrastructure planning and nature-based solutions. https://climate-adapt.eea.europa.eu/en/metadata/adaptation-options/green-spaces-and-corridors-in-urban-areas

Corpuz AG (2018) Infrastructure and urbanization. In: Clarete RL, Esguerra EF, Hill H (eds) The Philippine economy: no longer the East Asian exception? ISEAS-Yusof Ishak Institute, Singapore, pp 112–150

DNDA (n.d.) Delridge wetland and stewardship project. https://dnda.org/dnda-nature/wetlands/outdoorclassroom/. Accessed 29 June 2024

Do Y, Kim JY (2020) An assessment of the aesthetic value of protected wetlands based on a photo content and its metadata. Ecol Eng 150:105816. https://doi.org/10.1016/j.ecoleng.2020.105816

Dobbs R, Remes J, Manyika JM, Roxburgh C, Smit S, Schaer F (2012) Urban world: cities and the rise of the consuming class. McKinsey Global Institute

Ecoquartiers (2024) A new way of designing, building and managing the city sustainably Browse the interactive map of EcoDistricts [in French]. https://www.ecoquartiers.logement.gouv.fr/. Accessed June 2024

EEA (2021) NATURA Natura 2000 network viewer. European Environment Agency. https://natura2000.eea.europa.eu/. Accessed 29 June 2024

ENFIELD Council and ZSL (2021) Urban wetland design guide designing wetlands to improve water quality. Zoological Society of London, London

EU (2021) European commission: environment. Natura 2000. https://ec.europa.eu/environment/nature/natura2000/. Accessed 29 June 2024

European Commission Directorate-General for Environment (2014) Building a green infrastructure for Europe. https://doi.org/10.2779/54125 Accessed 29 June 2024

Gaber RM (2020) Implementation of constructed wetlands landscape design—a resilient application. J Urban Res 36(1):82–101. https://doi.org/10.21608/jur.2020.90104

Gulati M, Scholtz L (2020) The case for investment in green infrastructure in African Cities. Cape Town, WWF South Africa

Haberl R, Perfler R, Mayer H (1995) Constructed wetlands in Europe. Water Sci Technol 32(3):305–315. https://doi.org/10.1016/0273-1223(95)00631-1

Haron A, Feisal Z (2020) Constructed wetland parks: a pathway to sustainability for Cairo. Journal of Urban Research, 36(1), 61–81, https://doi.org/10.21608/jur.2020.90103

Lachinani Kordesofla M, Abna P (2023) Urban constructed wetlands in arid and semiarid zones. Civil Environ Eng 19(1):364–379. https://doi.org/10.2478/cee-2023-0033

Lavrnić S, Zapater Pereyra M, Cristino S, Cupido D, Lucchese G, Pascale MR, Toscano A, Mancini M (2020) The potential role of hybrid constructed wetlands treating university wastewater—experience from northern Italy. Sustain For 12(24):10604. https://doi.org/10.3390/su122410604

Liquete C, Udias A, Conte G, Grizzetti B, Masi F (2016) Integrated valuation of a nature-based solution for water pollution control. Highlighting hidden benefits. Ecosyst Serv 22:392–401. https://doi.org/10.1016/j.ecoser.2016.09.011

Makopondo ROB, Rotich LK, Kamau CG (2020) Potential use and challenges of constructed wetlands for wastewater treatment and conservation in game lodges and resorts in Kenya. Sci World J 2020:1–9. https://doi.org/10.1155/2020/9184192

Masi F, Rizzo A, Bresciani R, Conte G (2017) Constructed wetlands for combined sewer overflow treatment: ecosystem services at Gorla Maggiore, Italy. Ecol Eng 98:427–438. https://doi.org/10.1016/j.ecoleng.2016.03.043

Navarro M, D'Agostino A, Neri L (2020) The effect of urbanization on subjective well-being: explaining cross-regional differences. Socio Econ Plan Sci 71:100824. https://doi.org/10.1016/j.seps.2020.100824

Omayer HM, Haron A (2022) Stakeholders' visual preferences relationships of landscape design elements in constructed wetland parks. https://doi.org/10.21203/rs.3.rs-1458315/v1

Oppla (2023) Constructed wetlands as a multipurpose green infrastructure in Gorla Maggiore, Italy. https://oppla.eu/casestudy/17252. Accessed 29 June 2024

Orestad (2024) The Water Guild in Ørestad [in Danish]. https://orestad.net/kanaler/. Accessed June 2024

Pedersen E, Weisner SEB, Johansson M (2019) Wetland areas' direct contributions to residents' well-being entitle them to high cultural ecosystem values. Sci Total Environ 646:1315–1326. https://doi.org/10.1016/j.scitotenv.2018.07.236

Rizzo A, Conte G, Masi F (2021) Adjusted unit value transfer as a tool for raising awareness on ecosystem services provided by constructed wetlands for water pollution control: an italian case study. Int J Environ Res Public Health 18(4):1531. https://doi.org/10.3390/ijerph18041531

Ramsar Convention Secretariat (2013) The Ramsar convention manual: a guide to the convention on wetlands (Ramsar, Iran, 1971), 6th edn. Ramsar Convention Secretariat, Gland

UN (2019) World urbanization prospects: the 2018 revision (ST/ESA/SER.A/420) (P. D. United Nations, Department of Economic and Social Affairs, Ed.; Vol. 126). New York, United Nations Publications

UN-Habitat (2008) Constructed wetlands manual. United Nations Human Settlements Programme (UN-HABITAT)

US Environmental Protection Agency (2024) Overcoming barriers to green infrastructure. https://www.epa.gov/green-infrastructure/overcoming-barriers-green-infrastructure. Accessed 22 April 2024

Vargas-Hernández JG, Zdunek-Wielgołaska J (2020) Urban green infrastructure as a tool for controlling the resilience of urban sprawl. Environ Dev Sustain 1–20:1335–1354. https://doi.org/10.1007/s10668-020-00623-2

Velasco P, Devanadera MC, Dalisay M, Mueca C, Estorba DS, Lecciones A (2023) Nature-based solutions for domestic wastewater treatment in The Philippines. In: Regional perspectives of nature-based solutions for water: benefits and challenges. Springer, pp 175–201. https://doi.org/10.1007/978-3-031-18412-3_7

Wang W, Wu T, Li Y, Xie S, Han B, Zheng H, Ouyang Z (2020) Urbanization impacts on natural habitat and ecosystem services in the guangdong-hong kong-macao "megacity". Sustain For 12(16):6675. https://doi.org/10.3390/su12166675

WHO (2012) WHOQOL: measuring quality of life. https://www.who.int/tools/whoqol#:~:text=WHO%20defines%20Quality%20of%20Life,%2C%20expectations%2C%20standards%20and%20concerns. Accessed 29 June 2024

Zeng Y, Deal B (2023) What role do urban parks play in forming a sense of place? Lessons for geodesign using social media. Land 12(11):1960. https://doi.org/10.3390/land12111960

Zhang L, Schwärzel K (2017) Applying multifunctionality to address the challenges and benefits of land-use management. In: Zhang L, Schwärzel K (eds) Applying multifunctionality to address the challenges and benefits of land-use management, 1st edn. Springer AG, pp 1–9. https://doi.org/10.1007/978-3-319-54957-6

Zhu J, Wang Y (2022) The value of urban wetland parks and suggestions for development countermeasures. Open J Soc Sci 10(5):345–350. https://doi.org/10.4236/jss.2022.105022

Index

A. Battisti et al. (eds.), *Urban-Rural Dialogue in Green City Design*, Designing Environments, https://doi.org/10.1007/978-3-031-88318-7

The manufacturer's authorised representative in the EU is Springer Nature Customer Service Centre GmbH, Europaplatz 3, 69115 Heidelberg, Germany. If you have any concerns regarding our products, please contact ProductSafety@springernature.com

Printed and bound by CPI Group (UK) Ltd, Croydon, CR0 4YY
01/10/2026
02225949-0002